# Arduino

# Robotics

# Projects

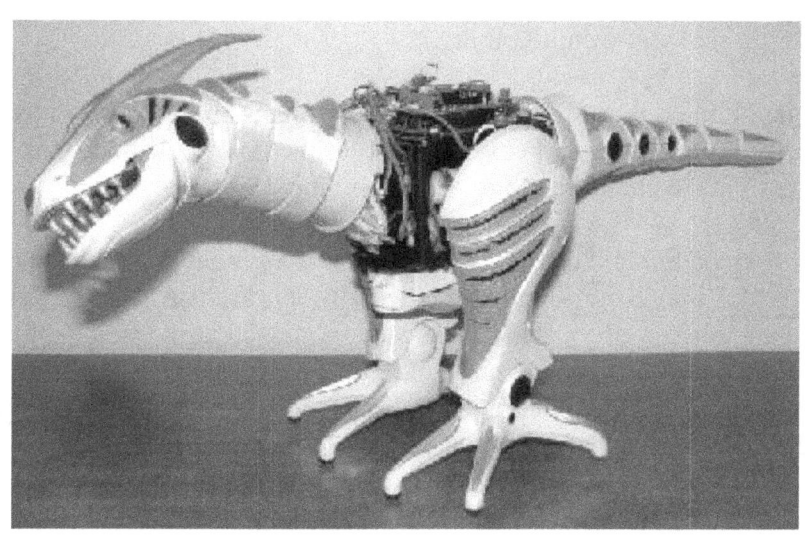

## Robert J Davis II

Arduino Robotics Projects
By Robert J Davis II
Copyright 2013 Robert J Davis
All rights reserved

As always, I have made every effort to make the projects in this book as simple, easy to build and as easy to understand as possible. However the reader or builder must take all safety precautions in building these projects. The safe construction and operation of these devices are solely the responsibility of the reader or builder. The author assumes no liability for the safe construction and operation of these devices.

This is the third book of my "Arduino projects" books. The first two were "Arduino LED Projects" and "Arduino LCD projects". Each book gets more technical and complicated. Each book builds on the knowledge that is gained in the previous books. It is highly recommended that you progress through the books in the order that they were written. However, if you are already well trained in electronics, you can join in wherever you want with whatever project you want to build.

Once again I had a lot of fun building these projects and I want you to also have fun in building them. Feel free to experiment with the code and hardware to make even more interesting and fun devices.

Robotics can be fun and exciting. Robots are used today to make creatures "come to life" in museums and in movies. Robots are used to help assemble cars and appliances. Robots are even used to make a farm tractors put the correct amount of fertilizer in the right place in a field via GPS. Robots are used to examine underground pipes and wires for defects. Robots are used to go places that humans cannot go, or that are to dangerous for us to go. Robots are used for just about anything else that you can imagine.

To understand robotics you need to understand motors, steppers, servos, valves, etc. We will look at these devices, see how they work, do some experiments with them and then we will apply what we have learned to making actual working robotic devices. We will start with building simple things like a remote control car and then advance up to making more complex robotic devices. Most of all, we will have lots of fun building and playing with some robotic devices.

Table of Contents:

1. DC Motors....................................................... 4
    DC Motor Control "H" or "Bridge"
    Dual Motor Control with L293 IC

2. Stepper Motors….............…....................…....11
    Transistor Stepper Drive
    ULN2803 Stepper Drive

3. Servo Motors.................................................. 22

4. Relays and Solenoids........................................ 27
    Conventional Relays
    Solid State Relays
    ULN2003 Relay Driver

5. Arduino Controlled Two Motor Robot...................... 31
    Octobot with IR proximity detectors

6. Arduino Controlled Toy Car................................ 35
    Toy Car with Ultrasonic proximity detector
    Toy Car with Wireless Bluetooth control
    Adding a Wireless CCTV Camera

7. Arduino Controlled Roomba Robot.......................... 54
    Roomba with Serial Communications
    Roomba with Ultrasonic demo program
    Roomba with IR Remote control
    Roomba with L298 motor driver IC

8. Arduino Controlled Homemade car.......................... 67
    Making the body of the car

9. Arduino Controlled RoboRaptor........................... 73
    Adafruit motor controller shield
    Raptor Demo Program
    Raptor with Serial Communications
    Raptor with IR Remote control

Bibliography.....................................................102

# Chapter 1

# DC Motors

You are likely familiar with the common small DC motors that are used in many children's toys. By applying about six volts DC to them you can make the motor spin. By reversing the power connections, you can then make the motor spin in the opposite direction.

Here is a picture of some typical motors that are fond in toys.

The typical standard DC motor has two electromagnets and two fixed magnets. When power is applied to the motor the electromagnets are turned on. By the rule of magnetism that states that opposites attract, the electromagnets move toward their opposing magnets. Just as the spinning electromagnet arrives at its destination, a set of brushes or contacts reverses power to the electromagnet, so it now continues spinning towards the other magnet. This process takes place so rapidly that all we see is the spinning motor.

You can not just connect an Arduino digital output pin directly to a DC motor. If you try to do that, you will likely damage the Arduino. The output of the Arduino is rated at 40 Ma, which is only .04 amps. DC Motors require a lot more power than that. So we will need to add a motor driver circuit to properly control a motor.

First, I want to explain what a motor control circuit is, and how it works. Basically you have two sets of NPN and PNP transistors that are configured to look like an "H" or a "Bridge". Newer controllers use "Field Effect Transistors" (FET's). When given a logic "one" or 5 volts, one transistor on one side of the motor conducts to apply power. At the same time a logic "zero" or ground on the other transistor on the other side of the motor conducts to apply ground. To reverse the direction the motor spins in, reverse the sides that are applying power and ground to the motor. If both sides apply power or ground at the same time, then the motor will not move.

Here is a typical schematic of a DC motor controller. The 6 volt power source needs to come from the "Vin" pin of the Arduino that is located near the A0 pin, or from an external power source. If not, then the electrical noise from the motor spinning can cause communication issues with the Arduino. The motor controller inputs, marked "in" in the schematic, go to D9 and D10 for the demo sketch listed below to work properly.

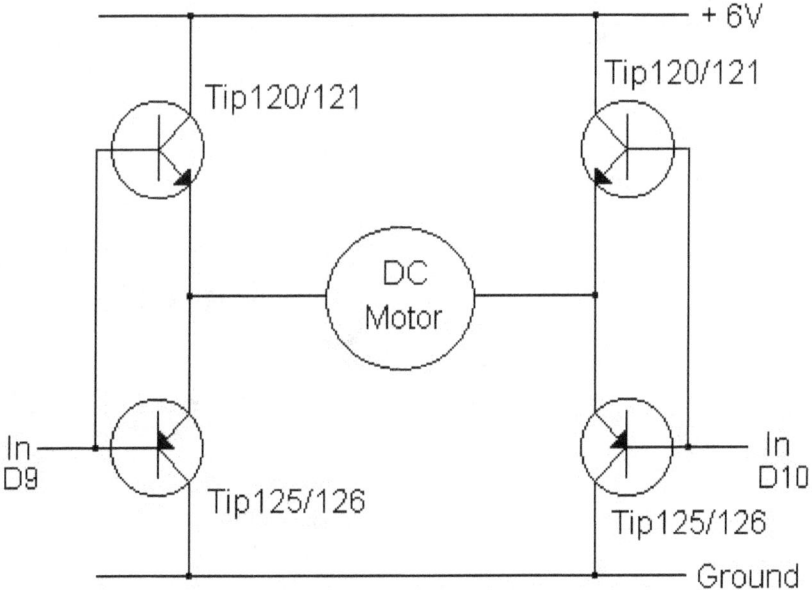

Below is a picture of a circuit board from a toy car. You can easily see that there are three "H" Bridges by looking at the arrangement of the transistors on the board. The two on the right side are higher power versions. The one on the left side is a lower power version.

Here is a picture of a motor controller "H" configuration built on a breadboard. Note that I mounted the transistors so they are turned sideways. That way the wider leads of the power transistors do not cause damage to the breadboard. Also the pin names on the transistors are from the left to the right Base, Collector, and Emitter. The Base is the input, the collector is the output without an arrow, and the emitter is the output pin with an arrow in the schematic diagram that was shown above.

Another thing that you need to be aware of is that some of the digital outputs of the Arduino can be more than just "on" and "off". Some of the digital outputs can be "Pulse Width Modulated" (PWM), to have what is essentially a voltage between zero and five volts depending upon the width of the pulse. Sometimes theses are referred to as analog outputs. By using these pins we can control not only the direction of the motor, but also the speed of the motor. To do all of that, we only need two digital output pins of the Arduino that support the PWM modulation mode.

Here is a sketch to make the motor controller work. The center of a variable resistor is connected to the A0 pin with the resistors ends connected to ground and five volts. This variable resistor then varies the motor from spinning in one direction, to sitting still, to spinning in the other direction as you move its knob.

```
/***************************
Motor controller demo
by Bob Davis
July 5, 2013
***************************/
int motor1 =9;
int motor2 =10;
int mspeed =0;

void setup() {
pinMode(motor1, OUTPUT);
pinMode(motor2, OUTPUT);
}

void loop() {
// analog read is 0 to 1024, analog write is 0 to 255
mspeed=analogRead(A0)/2;
if (mspeed < 256) {
// vary the ground via PWM
  analogWrite (motor1, mspeed);
// turn on power output
  digitalWrite(motor2, HIGH);
}
if (mspeed > 256) {
// vary the ground via PWM
  analogWrite (motor2, 255-mspeed);
// turn on power output
```

digitalWrite(motor1, HIGH);
}
}

Of course, some companies have made some motor controller Integrated Circuits or "IC's". A very popular motor controller IC is the L293D. It contains two of the H or bridge motor drivers in one IC. Here is the L293D's connection or schematic diagram. There is even an Arduino "shield" available that has two of the L293 motor controllers mounted on it. The L293 has over current protection built into it to protect the controller from short circuits.

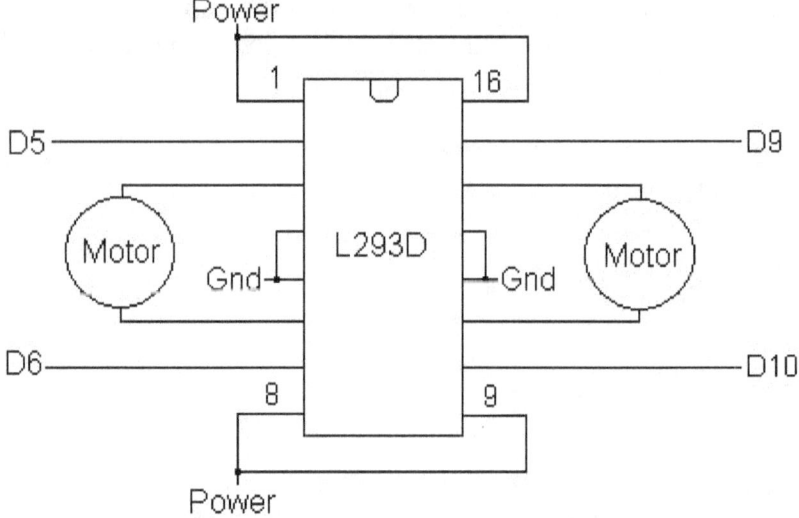

Here is a picture of the L293 based dual motor controller. The sound of the two motors running together sounds something like an airplane running.

Here is a sketch that was written to control the two motors with the L293 IC. The centers of two variable resistors are connected to A0 and A1 to control the speed of the motors. The other ends of the variable resistors go to five volts and ground.

```
/***************************
Motor controller demo
by Bob Davis
July 5, 2013
***************************/
int motor1A =5;
int motor1B =6;
int motor2A =9;
int motor2B =10;
int mspeed =0;

void setup() {
pinMode(motor1A, OUTPUT);
pinMode(motor1B, OUTPUT);
pinMode(motor2A, OUTPUT);
pinMode(motor2B, OUTPUT);
}

void loop() {
// analog read is 0 to 1024, analog write is 0 to 255
mspeed=analogRead(A0)/2;
if (mspeed < 256) {
// vary the ground via PWM
  analogWrite (motor1A, mspeed);
// turn on power output
  digitalWrite(motor1B, HIGH);
}
if (mspeed > 256) {
// vary the ground via PWM
  analogWrite (motor1B, 255-mspeed);
// turn on power output
  digitalWrite(motor1A, HIGH);
}

// Motor 2
mspeed=analogRead(A1)/2;
```

```
if (mspeed < 256) {
// vary the ground via PWM
  analogWrite (motor2A, mspeed);
// turn on power output
  digitalWrite(motor2B, HIGH);
}
if (mspeed > 256) {
// vary the ground via PWM
  analogWrite (motor2B, 255-mspeed);
// turn on power output
  digitalWrite(motor2A, HIGH);
}
}
```

# Chapter 2

# Stepper Motors

With a normal motor it is very difficult to rotate the motor to a specific, exact position. Stepper motors can solve that problem of conventional motors, in that they can be rotated to any specific position. Rotating to a specific position is done by having many electromagnets and a rotating iron gear. A typical four phase stepper motor has four, eight, or more sets of coils and a rotating iron gear. As each coil, or sets of opposite coils, is energized the gear rotates to align with the active coils. Most stepper motors have four phases and five wires. One wire is for each of the four sets of coils and a common wire is for the power source.

Here is what the guts of a typical stepper motor looks like with one end of the motor removed so you can see inside of it. You can see five of the white coil forms located around the blue colored rotating iron gear. It looks like the brown wire might be the common, or power wire and the other four wires are for the four phases.

11

The typical single phase stepping motor control signals might look something like this chart:

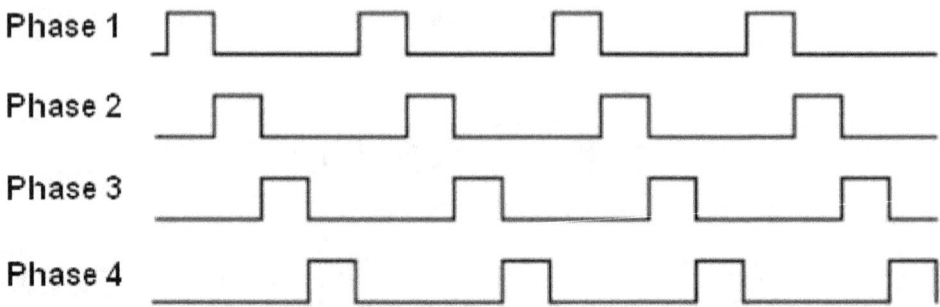

Basically, as each phase fires, the motor rotates to that position. However there is a problem with this method of stepping the motor. As one phase turns off, and then another phase turns on, there is a "dead" time. This dead time makes the stepper motor "weak". The stepper motor will have a hard time holding the correct position. One solution is to have "overlapping" phases or steps. That is to say that one phase overlaps with another phase. This is called "two phase", or "full step" stepper motor drive.

Here is what two phase stepping looks like.

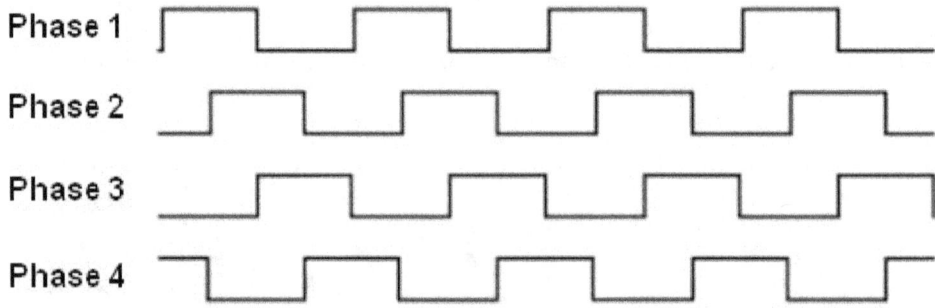

In two phase stepping two motor phases are on at any one time. That way there is no "dead" time between phases and the motor is always locked to a specific position. With two phases on, the stepper motor has twice the power as when using single phase stepping.

Another stepper motor technique is called "half stepping". Years ago it was used by Apple and others to copy protect their floppy disks. They half stepped and wrote data that was essentially "between the tracks" on a floppy disk. That way no one could copy the data on the floppy disk, unless they knew where the half track data was stored.

Here is a chart showing what "half stepping" looks like:

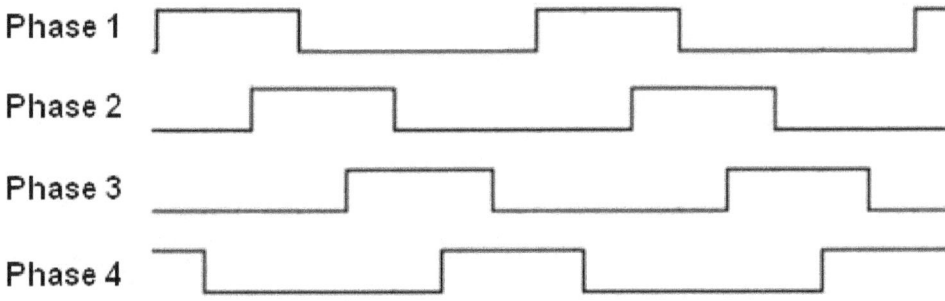

Basically you have phase one that is on by itself, then phases one and two are on together. Then phase two is on by itself, and then phases two and three are on together. Next phase three is on by itself, and then phases three and four are on together. Finally, phase four is on by itself, and then phases four and one are on together. This half stepping mode replaces the four steps with eight steps giving more positions or steps. This mode also works with no dead time because one phase is always active.

With a stepper motor, and these stepping techniques, you can position the motor very precisely. Stepper motors were used to position the heads on hard drives and floppy drives for many years, until there were too many tracks for stepper motors to select all of them. They then switched to a "voice coil" motor that offers an infinite number of steps. Typically, in a normal stepper motor, you have steps of 1.8 degrees of rotation or 200 steps per 360 degrees of revolution. Adding half stepping gives you steps of .9 degrees of rotation or 400 steps per revolution. Stepper motors are still used in printers to position the print head and the paper today.

Now, how do you connect a stepper motor to an Arduino? Once again you cannot just connect the motor to the Arduino output pins. You will need to add a driver transistor for each of the four phases.

Determining the color code of the wires for a stepper motor can be fun to do. Basically you need to determine what motor pin is connected to power, by using a voltmeter. Then, the four phases can be "played with" until you get a smoothly stepping motor. If the stepper motor jumps around the phases are out of order.

Below there is a typical schematic of a stepper motor driver setup using power transistors. By the way, the 12V comes from the "Vin" pin on the

Arduino and it can be anything from 6 to 12 volts DC. It can also be an external power source.

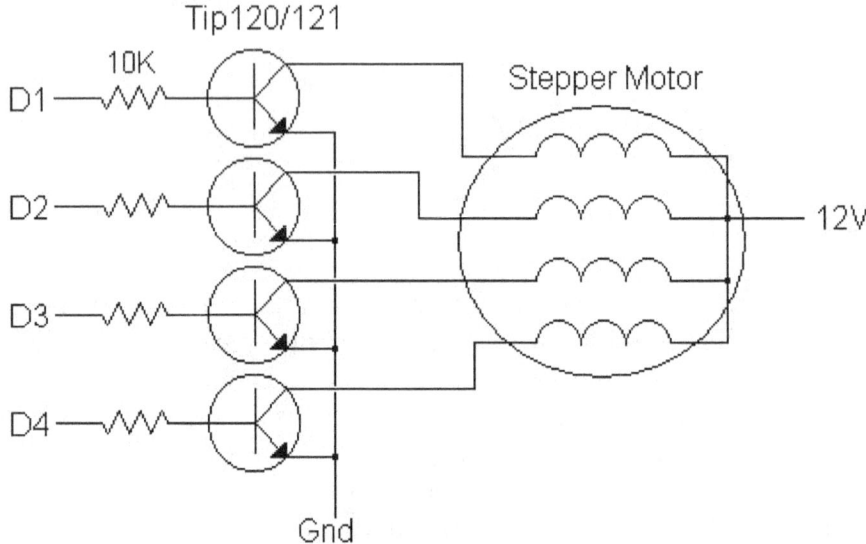

Up next is a picture of a working stepper motor controller using individual driver transistors for each phase assembled on a breadboard.

Here is the sketch or code to make the stepper motor rotate at a speed that depends upon the position of a variable resistor with the variable resistors center connected to A0. The other two ends of the variable resistor go to ground and to five volts.

```
/****************************
Stepper Motor controller demo
by Bob Davis
July 5, 2013
****************************/
int motor1 = 1;
int motor2 = 2;
int motor3 = 3;
int motor4 = 4;
int mspeed = 0;
int phase = 0;
//  allow redefining of pins
int pinoffset = motor1-1;

void setup() {
pinMode(motor1, OUTPUT);
pinMode(motor2, OUTPUT);
pinMode(motor3, OUTPUT);
pinMode(motor4, OUTPUT);
}

void loop() {
// analog read is 0 to 1024, that is very slow
mspeed=analogRead(A0);
for (phase=1; phase <5; phase++) {
  digitalWrite(phase+pinoffset, HIGH);
  if (phase < 4) digitalWrite(phase+1+pinoffset, HIGH);
  if (phase == 4) digitalWrite(1+pinoffset, HIGH);
// a delay of < 5 is too fast for the stepper motor
  delay(mspeed + 5);
  digitalWrite(phase+pinoffset, LOW);
}}
```

Here is a sketch to make the motor rotate to match the position of a variable resistor that is connected to A0.

```
/***************************
Stepper Motor positional Controller demo
by Bob Davis
July 5, 2013
***************************/
int motor1 = 1;
int motor2 = 2;
int motor3 = 3;
int motor4 = 4;
int mposit = 0;
int oldposit = 0;
int cposit = 0;
int phase = 0;

// allow redefining of pins
int pinoffset = motor1-1;
void setup() {
pinMode(motor1, OUTPUT);
pinMode(motor2, OUTPUT);
pinMode(motor3, OUTPUT);
pinMode(motor4, OUTPUT);
}

void loop() {
// analog read / 20.48, leaves 50 x 4 phases
// = 200 steps or 1 complete revolutions
mposit = analogRead(A0)/20.48;
cposit = mposit-oldposit;
oldposit = mposit;
while (cposit != 0) {
// Process positive numbers by rotating clockwise
  if (cposit > 0) {
    for (phase=1; phase <5; phase++) {
      digitalWrite(phase+pinoffset, HIGH);
      if (phase < 4) digitalWrite(phase+1+pinoffset, HIGH);
      if (phase == 4) digitalWrite(1+pinoffset, HIGH);
      // a delay of < 5 is too fast for the stepper motor
      delay(10);
      digitalWrite(phase+pinoffset, LOW);
    }
    cposit--;
  }
```

```
// Process negative numbers by rotating counter clockwise
  if (cposit < 0) {
    for (phase=4; phase >0; phase--) {
      digitalWrite(phase+pinoffset, HIGH);
      if (phase > 1) digitalWrite(phase-1+pinoffset, HIGH);
      if (phase == 1) digitalWrite(4+pinoffset, HIGH);

      // a delay of < 5 is too fast for the stepper motor
      delay(10);
      digitalWrite(phase+pinoffset, LOW);
    }
    cposit++;
  }
}
}
```

Of course once again they make an IC that can do the job of running a stepper motor. This IC can not only run one stepper motor but it can drive two stepper motors. The IC is the ULN2803 and it contains eight common emitter NPN Darlington power transistors. Here is a typical ULN2803 schematic diagram.

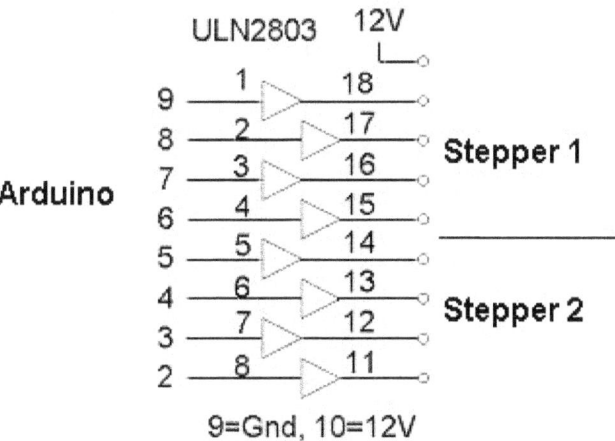

The ULN2803 has eight power Darlington transistors, so is a very universal IC. You will find it used to operate relays, stepper motors, conventional motors, solenoids, LED signs, and almost anything that needs to be turned on and off with more power than what a normal IC can handle. As you can guess the IC takes up a lot less room on a circuit board than the four power transistors did.

This next picture shows some typical stepper motors. The most common size is called NEMA 17, they are around 1.7 by 1.7 inches in size and can be seen in the front row. Another common size is called NEMA 23, they are around 2.3 by 2.3 inches in size, as seen in the back row.

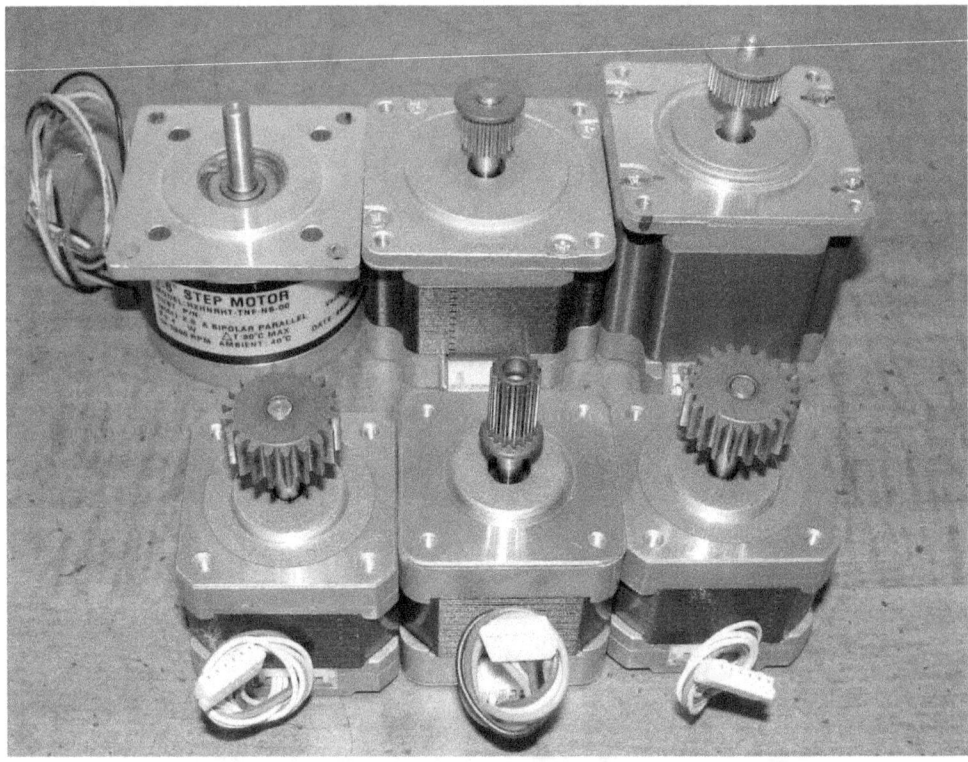

Stepper motors are also available in configurations other than the five wire version. The two most common other configurations are the four and six wire versions. The four wire version requires two "H" bridges or an L293D IC to operate. The six wire stepper motor can either be configured as a four wire stepper motor, by ignoring the center taps, or as a five wire stepper motor by connecting the two center taps together and to the power source. Here is a schematic of a typical six wire stepper motor.

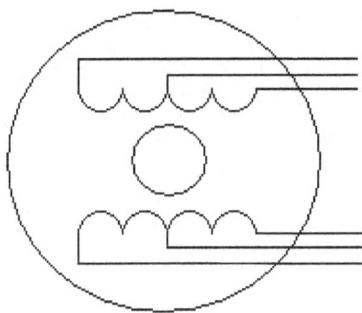

The next schematic shows how to wire up either a four wire or a six wire stepper motor using a L293D IC. I actually used this to setup to test several stepper motors. You can use Vin of the Arduino to power it, but an external 9 to 12 volt source will give the stepper motor a lot more power. If you reverse any of the connections to the stepper motor the motor will just spin in the opposite direction.

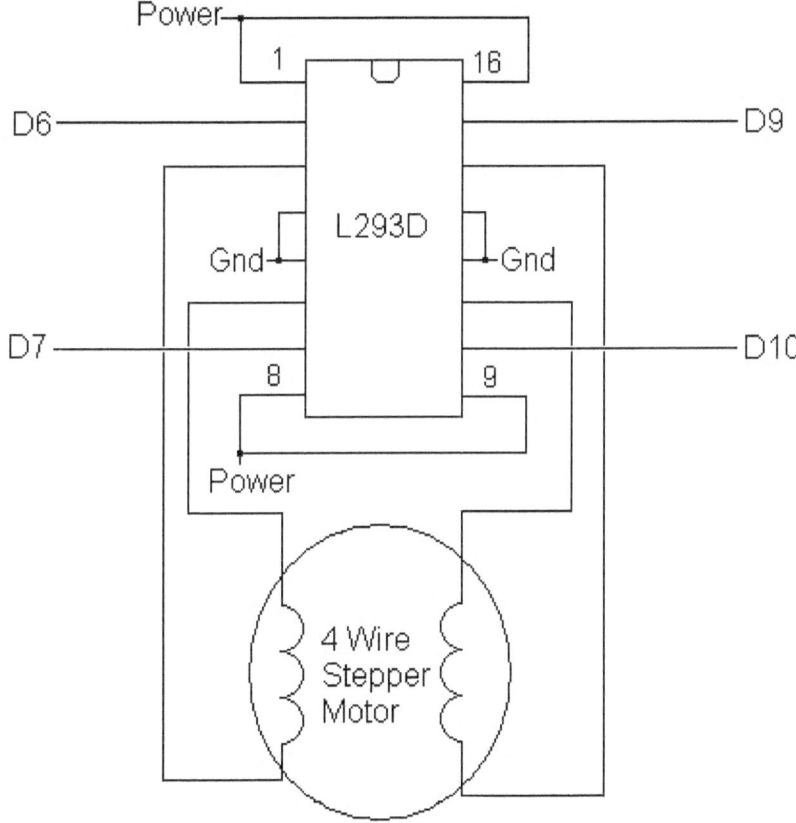

The next picture is of a L293D IC running a six wire stepper motor that is configured as a four wire stepper.

Here is a sketch to run a four wire stepper at a speed that depends upon the position of a variable resistor that is connected to A0.

```
/*****************************
Four wire stepper motor demo
by Bob Davis
February 5, 2014
*****************************/
int motor1A =6;
int motor1B =7;
int motor1C =9;
int motor1D =10;
int mspeed =0;

void setup() {
pinMode(motor1A, OUTPUT);
pinMode(motor1B, OUTPUT);
pinMode(motor1C, OUTPUT);
pinMode(motor1D, OUTPUT);
}

void loop() {
```

```
// analog read is from 0 to 1024
mspeed=(analogRead(A0)/5);
  digitalWrite (motor1A, HIGH);
  digitalWrite(motor1B, LOW);
  digitalWrite (motor1C, HIGH);
  digitalWrite(motor1D, LOW);
delay(mspeed);
  digitalWrite (motor1A, HIGH);
  digitalWrite(motor1B, LOW);
  digitalWrite (motor1C, LOW);
  digitalWrite(motor1D, HIGH);
delay(mspeed);
  digitalWrite (motor1A, LOW);
  digitalWrite(motor1B, HIGH);
  digitalWrite (motor1C, LOW);
  digitalWrite(motor1D, HIGH);
delay(mspeed);
  digitalWrite (motor1A, LOW);
  digitalWrite(motor1B, HIGH);
  digitalWrite (motor1C, HIGH);
  digitalWrite(motor1D, LOW);
delay(mspeed);
}
```

# Chapter 3

# Servo Motors

Servos are motors that have a built in motor controller and a "feedback" loop. Basically a variable resistor or other device monitors the motor's position. This information is then "fed back" to the built in motor controller. The motor controller takes commands from a processor, usually the commands are in the form of pulse width modulation, and then the controller matches up the motor position with what position the controller was told to move the motor to.

Servo's typically come in several sizes, as can be seen in the above picture. There are the normal sizes of servo's on the left and the micro sized servo on the right.

In the case of pulse width modulation commands, usually a pulse width of one millisecond tells the controller to move the servo to zero degrees. A pulse width of 1.5 milliseconds tells the controller to move the servo to 90 degrees. A pulse width of two milliseconds tells the controller to move the servo almost completely open or 180 degrees. The servo "home" position is at 90 degrees. In my second servo test program, using some cheap servo's found on eBay, I discovered that a .5 millisecond pulse results in zero degrees of rotation and a 2.5 millisecond pulse results in 180 degrees of rotation. In either case there is also a 20 millisecond delay between each of the control pulses.

This chart shows the pulse width and the corresponding position of the servo motor.

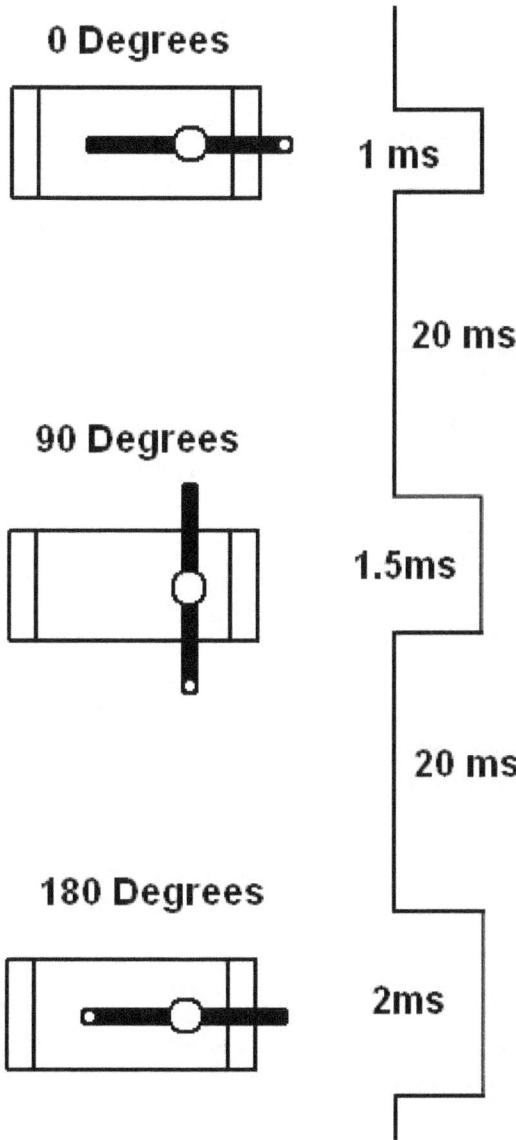

Some of the advantages of servo motors include that the power source does not have to be switched on or off or otherwise controlled. The power to the servo motor can always be left "on". This time a PWM output pin of the Arduino can directly control the servo, no driver circuit or transistor is needed, because there is a driver inside of the servo. Servos make great proportional valve controllers because you can vary a valve from off to full on. For instance, if you want to water your plants automatically and you want the rate of water flow to be adjusted according to the humidity, it can be done with a servo.

This is a picture of the Arduino servo motor test setup.

Here is a sketch to demonstrate the operation of a servo motor. Once again the output of a variable resistor is connected to A0. This demo uses the Arduino servo library.

```
/********************************
// Servo motor demonstration program
// By Bob Davis
// July 10, 2013
// Servo Connected to Gnd, Vin, and D9
// Variable resistor on AO, high end=5V and low end=Gnd
/********************************/
#include <Servo.h>
Servo demoservo;
```

```
// The variable resistor is on A0
int vrpin = 0;
int pos;

void setup()
{
// The servo is on pin 9
  demoservo.attach(9);
}

void loop()
{
// Read the variable resistor, 1024/5=205 degrees rotation
// Values over 180 are ignored
  pos = analogRead(vrpin)/5;
// send the position to the servo
  demoservo.write(pos);
  delay(25);
}
```

Here is a second servo demo sketch that does not use the Arduino servo library.

```
/********************************
// Servo motor demonstration program 2
// By Bob Davis
// July 10, 2013
// Servo Connected to Gnd, Vin, and D9
// Variable resistor on AO, high end=5V and low end=Gnd
// This demo does not use servo.h as the timing is included
/********************************/
// The variable resistor is on A0
int vrpin = 0;
int pos;
 void setup()
{
// The servo is on pin 9
  pinMode (9, OUTPUT);
}

void loop()
{
```

```
// Read the variable resistor, 1024*2 = 2048 = 2 milliseconds
// add 500 as .5 is the minimum milliseconds
  pos = (analogRead(vrpin)*2)+500;
// send the position to the servo
  digitalWrite (9, HIGH);
  delayMicroseconds(pos);
  digitalWrite (9, LOW);
  delay(20);
}
```

# Chapter 4

# Relays and Solenoids

When all else fails, another way to control things is with either "relays" or "solenoids". Relays can be used to control any voltage or current from less than one volt to thousands of volts and from less than one amp to thousands of amps. Basically a relay is an electromagnetic coil that then flips a switch or a set of switches. The switches can be single pole (one switch) or even four poles or more. Like relays, solenoids are also mechanical devices. An electromagnet pulls on something that in turn opens a valve or unlocks a door.

There is a type of relay that is called a "solid state" relay. Solid state relays work without having any moving parts to wear out. A "solid state" relay usually has an optical isolator. An optical isolator is a LED and a light sensor that provides over 1000 volts of isolation. Then the solid state relay has a Silicon Controlled Rectifier (SCR) or a Triac.

SCR's are diodes with a control connection to turn the diode on and off. A Triac is essentially two SCR's so it works both ways for controlling AC devices. The SCR in the solid state relay is what actually turns the power on and off. The optical isolator gives complete isolation between the input and the outputs. That makes solid state relays ideal for controlling 120, 220 or other high voltage devices with a low voltage input such as an output pin of the Arduino.

Here is a picture of some typical small common relays. In the top left there are some "plug in" types of relays. Being able to plug them in makes replacing them much easier. In the upper right corner there are a couple of typical sockets for the plug in relays. In the bottom left of the picture there are some solder in type of relays that are not much bigger than an IC. In the bottom right corner there are some "reed" relays. The contacts in a reed relay are in a vacuum so they are a more sensitive device.

Like other mechanical devices in this book you cannot directly drive a relay from an Arduino output pin. You should always use either a relay driver transistor or if you are driving more than two or three relays you should use an IC like the ULN2003. The transistor driver would be identical to the transistor drivers for the stepper motors. The ULN2003 IC is almost the same thing as the ULN2803 but it only has seven circuits instead of eight.

Coming up next is the ULN2003 relay driver schematic diagram. The pin marked 12 volts in the schematic goes to the power source for the relays and can be anything from five volts to 24 volts. There are reverse polarity protection diodes built into the ULN2003 that go to the power pin and should be connected to the relays power source to protect the driver IC from reverse polarity power spikes that commonly happen with relays.

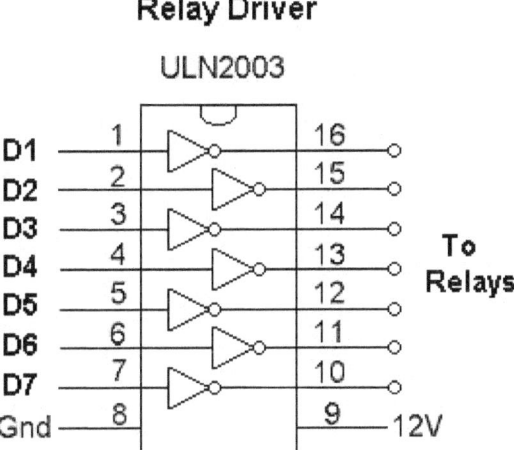

Relays can be configured to power on and to reverse the direction of a motor much like the transistors did in the "H" bridge configuration. This is a schematic of a relay motor control circuit configured in an "H" bridge setup. Each relay selects between power and ground going to one side of the motor.

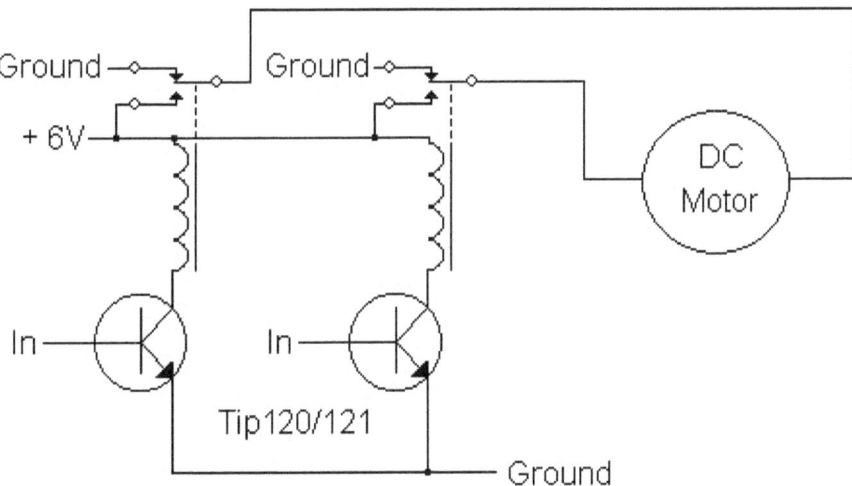

The most popular relay motor control setup is not an "H" bridge like what we used with the transistorized motor drivers. Relay motor controllers usually use two relays with one to turn the motor on and off and a second relay to select normal or reverse spin on the motor. The schematic usually looks something like this. Note that this setup requires that the relays be Double Pole Double Throw (DPDT) relays.

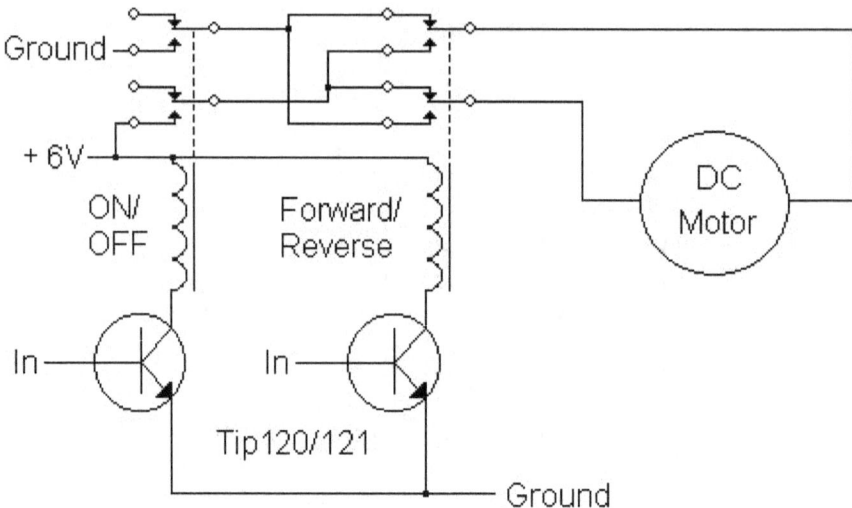

Relays can be found in a lot of strange places. A while back I had to install a 50 horsepower 440 volt three phase AC motor. It required a very large relay to switch it from "Star" to "Delta" configuration when the motor reached a certain speed.

# Chapter 5

# Arduino Controlled

# Two Motor Robot

Recently I was able to purchase what is called an "Octobot". It is a two wheel drive, eight sided, toy robot. The Octobot was designed to use a Basic Stamp controller, but it has a PIC controller that is actually running the motor for some reason. On my Octobot the "H" drive control for one of the motors had burned up. The drive controller had burned up because the gearbox was exposed underneath. That allowed it to swallow up into its gears almost anything that it passed overtop of. The exposed gear problem is easily fixed with a piece of tape over the opening.

The Octobot is just one example of a two motor robot. Basically you have two motors driving two wheels and then a caster or third wheel that follows along to keep the robot from tipping over. The software and hardware that are used in this project to run the Octobot would work for any other two motor robot.

I rebuilt the Octobot using an Arduino UNO and a L293 motor controller. The old PIC and Basic Stamp processors have to be removed to do this. The motor wires are not connected to the Octobot expansion connector. However the motor wires were just so long enough to reach a breadboard or small circuit board that was located towards the top rear of the Octobot. The Arduino goes in the center, above and between the wheels. Be sure to put tape over the metal binding posts. Only one mounting screw lined up with a hole in the Arduino.

Coming up next is the pinout listing for the Octobot's 20 pin expansion connector.

| Towards back | Pin Numbers | | Towards front |
|---|---|---|---|
| Ground | 1 | 2 | Ground |
| Left Photodetector | 3 | 4 | Left IR Proximity |
| Left IR Beacon | 5 | 6 | Right IR Beacon |
| | 7 | 8 | X2 Bus Pin2 |
| RIR Proximity | 9 | 10 | X2 Bus Pin 3 |
| | 11 | 12 | X2 Bus Pin 4 |
| Charger Cont. | 13 | 14 | X2 Bus Pin 5 |
| | 15 | 16 | Right Photodetector |
| PIC Serial in | 17 | 18 | PIC Serial out |
| +5V regulated | 19 | 20 | Battery |

We will need to tap into the Octobot expansion connector pins four and nine to monitor the IR Proximity detectors. Also tap into pin one for ground and pin 19 for regulated five volts to power the Arduino.

Here is the schematic of the Octobot's IR Proximity detector. Basically you have an oscillator that is connected to an IR LED that then sends out a pulsing IR light source. Then a Phase Locked Loop (PLL) looks for an IR echo back to see if it matches the transmitted frequency. The 10K resistor between the NE567 pins 5 and 6 sets the frequency. That resistor can be 6.8K, 8.2K, 10K or 12K so as to have different frequencies for each IR proximity detector to reduce their interaction with each other.

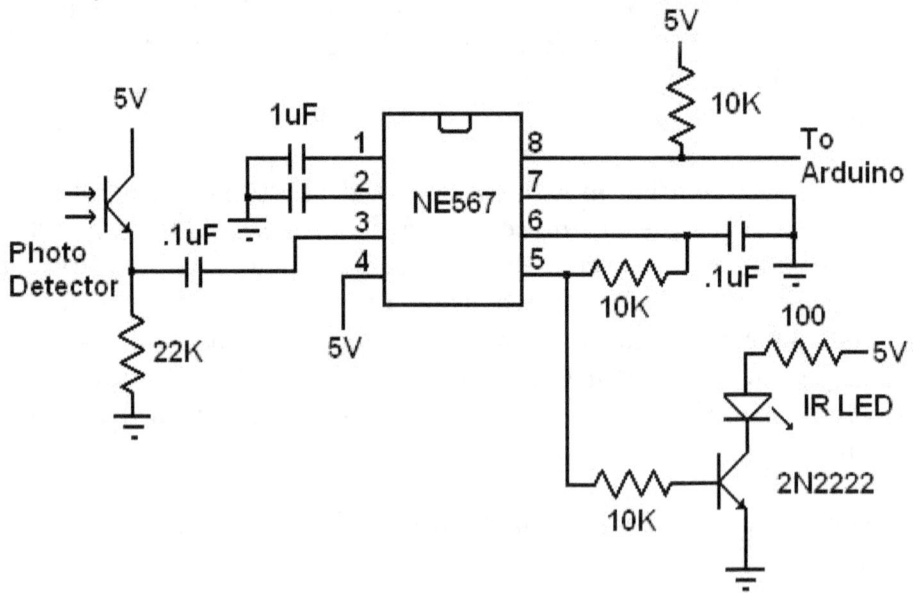

Up next is a picture of the Arduino powered Octobot robot. It shows the Arduino and motor controller breadboard mounted on top of the Octobot. Be sure to insulate with some electrical tape underneath of the Arduino.

Here is the sketch to make the Arduino powered Octobot work.

/*****************************
Two wheel controller demo
by Bob Davis
July 5, 2013
*****************************/

int motor1A = 5;
int motor1B = 6;
int motor2A = 9;
int motor2B = 10;
int prox1 = 0;

```
int prox2 = 0;

void freerun() {
  analogWrite (motor1B, LOW);
  digitalWrite(motor1A, HIGH);
  analogWrite (motor2B, LOW);
  digitalWrite(motor2A, HIGH);
}

void turnleft() {
  analogWrite (motor1A, LOW);
  digitalWrite(motor1B, HIGH);
  delay(100);
}

void turnright() {
  analogWrite (motor2A, LOW);
  digitalWrite(motor2B, HIGH);
  delay(100);
}

void setup() {
  pinMode(motor1A, OUTPUT);
  pinMode(motor1B, OUTPUT);
  pinMode(motor2A, OUTPUT);
  pinMode(motor2B, OUTPUT);
}

void loop() {
// analog read of prox sensors
prox1 = analogRead(A0);
prox2 = analogRead(A1);
// collision likely
if (prox1 < 100) turnleft();
if (prox2 < 100) turnright();
freerun();
}
```

# Chapter 6

# Arduino Controlled Toy Car

For this next project we are going to combine the two line LCD 1602 display from the LCD projects book, with the L293 two motor controller IC. Then we will add two HC-SR04 ultrasonic range finders. Ultrasonic range finders are much more accurate than the IR devices that we used on the robot in the previous project. IR devices use reflected light and the amount reflected light varies with the size and color of the objects surface. White is the best color for reflecting IR light, and black reflects the least amount of light.

Ultrasonic range finders use echoed high frequency sound similar to what is used for sonar on a submarine or what a bat uses in flight. Ultrasonic detectors can detect an object that is several feet away and even tell you almost exactly how far it is to that object.

Here is a specification sheet on the HC-SR04 ultrasonic detectors.

| Parameter | HC-SR04 ultrasonic module |
|---|---|
| Operating voltage | DC 5V |
| Operating current | 15 ma |
| Operating frequency | 40 KHz |
| Longest range | 4 m |
| Closest range | 2cm |
| Measurement angel | 15° |
| Input trigger signal | 10uS TTL pulse |
| Output echo signal | Output high TTL level signal, The length is proportional to the range |
| Physical Size | 45*20*15 mm |

Here is a picture of the two line LCD setup from my LCD projects book. For this project I soldered jumpers from pin one to pin five and to pin 16 and then another jumper from pin two to pin 15. I also hard soldered the contrast trimmer resistor on the back side of the LCD as you can see in the picture.

That way all that is needed to connect the LCD up to the Arduino is a 10 conductor male to female jumper cable.

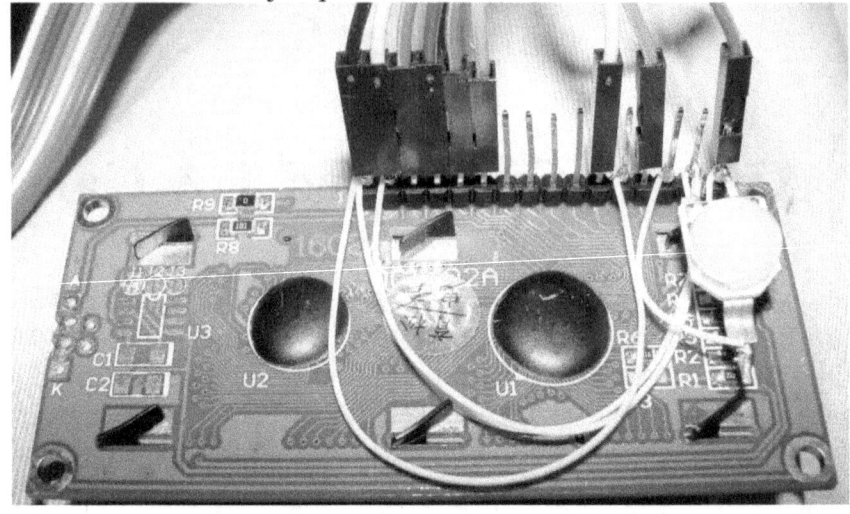

Here is the schematic for the 1602 LCD. It is from my "Arduino LCD Projects" book. It shows how to connect a two line LCD to the Arduino.

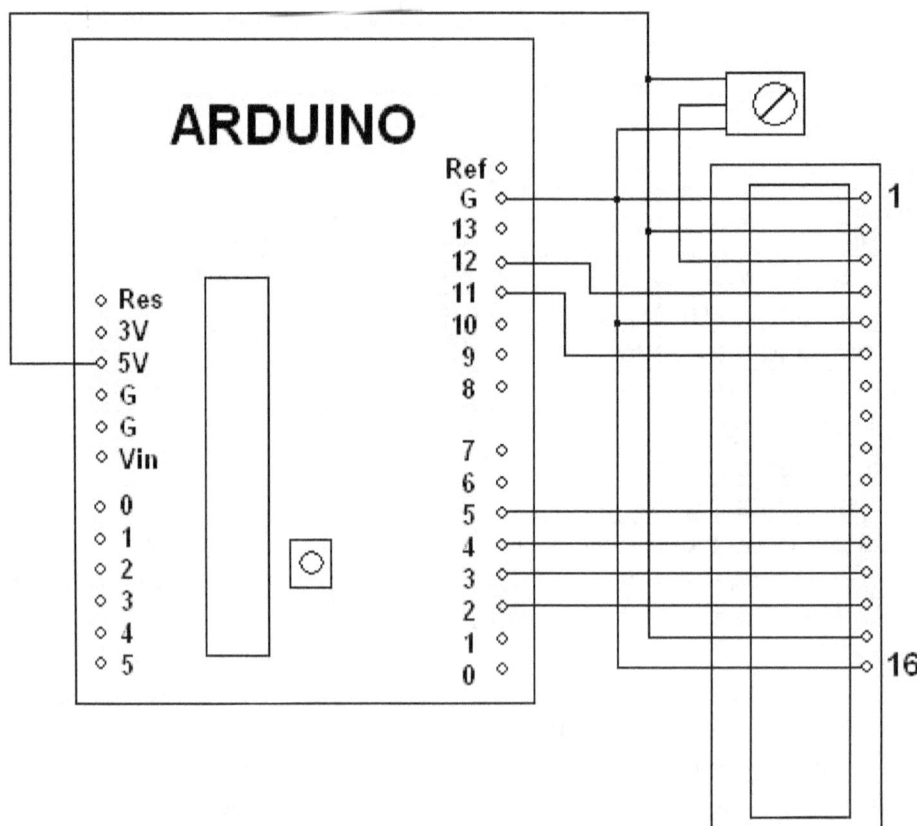

Here is the L293 motor controller; it has been modified from the one that is found earlier in this book. I converted it from the breadboard version to a "shield" version to increase the overall reliability of the circuit. I added a six pin header connector to the +5V and Ground pins on the left side of the circuit board. I did that to resolve an eventual shortage of power and ground pins.

On the bottom left is the external power connector. Power for the motor controller can come from the Vin Arduino pin via a red wire. Later on I also added a 220 uF 10 volt capacitor across the Vin pin to Ground to improve reliability. The four connectors with the blue wires on the right go to the motors.

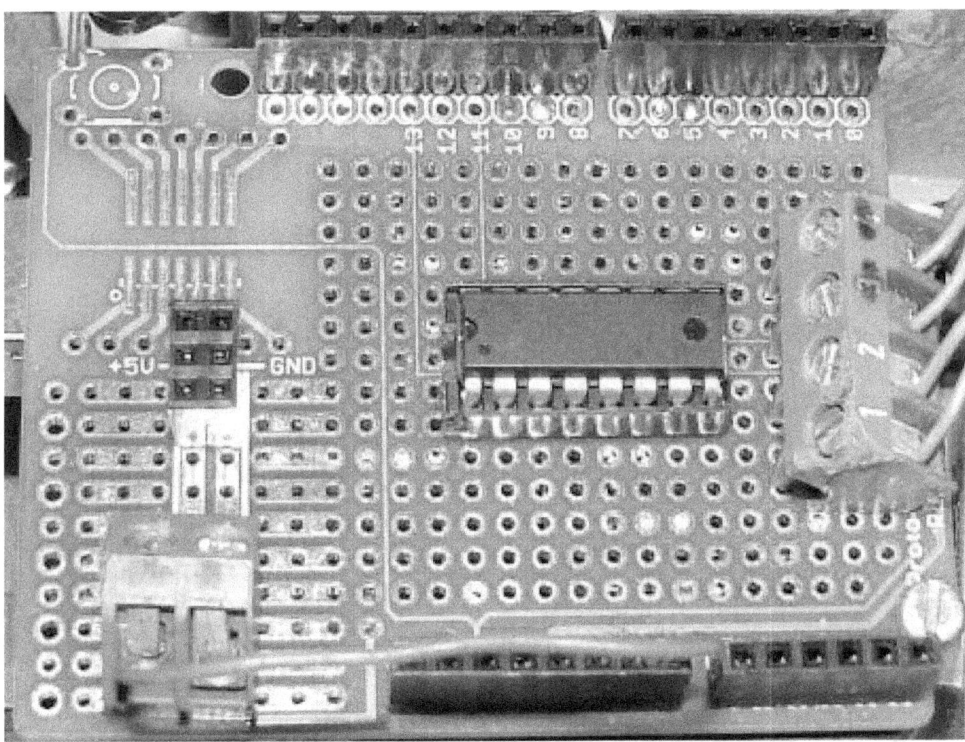

On the next page there is the motor controller schematic diagram that was given earlier in this book. Note that for this project, there was a pin usage conflict because pin D5 in the earlier schematic is now being used by the LCD display. That pin had to be changed so that it is connected to pin D7 instead of D5 for this project to work properly.

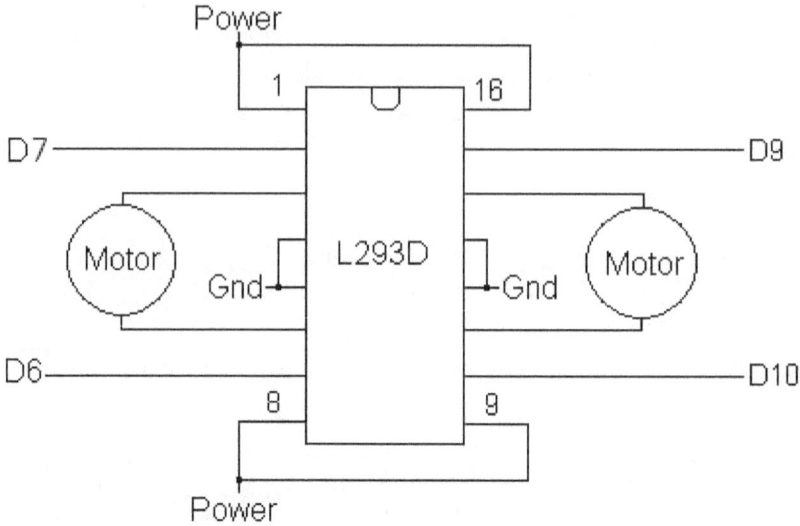

Here is what the two HC-SR04 ultrasonic detectors look like. They only have four pins to connect them up with. From left to right the four pins are power, trigger, echo and ground.

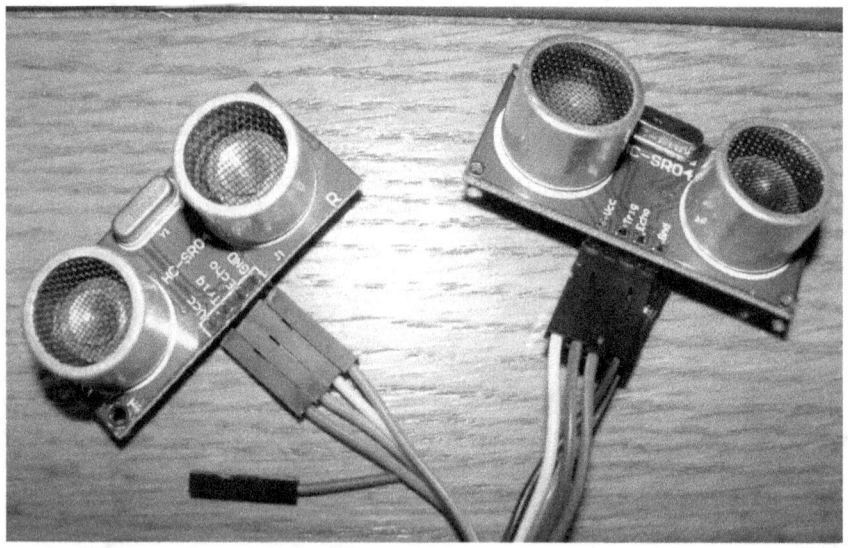

Here is the wiring schematic of the two Ultrasonic detectors. Note that the Arduino pins that are numbered D14 to D17 are the same as the A0-A3 pins. The analog pins are just being used as digital pins instead of as analog pins.

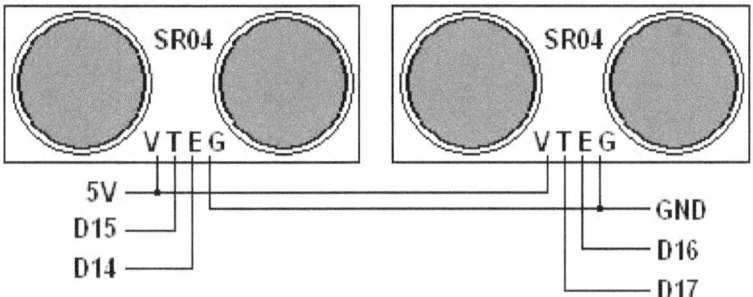

This is the front view of the toy car with the ultrasonic distance sensors in place. They were held in place with hot melt glue. They should have been mounted a little higher up as there was some interference with the range detection coming from the engine hood of the car.

The next picture is the view of the back of the car with the two line LCD display mounted in the back window.

The next picture will show what the toy car's internal wiring looked like with the top of the car removed. At some point I also added a nine volt battery to power the Arduino. I did not have a battery clip adapter to connect to the power jack, so the battery wires were coated in solder and then pushed into the Arduino Vin and Ground positions.

When the Arduino was powered by using the same four AA cells as the motor controller, the car had a tendency to reset the power to the Arduino whenever the motor reversed direction. To prevent the resetting problem you need to separate the power to the motors. To do that, connect the six volt lead from the four AA batteries into the motor controller board. Then use a separate nine volt battery to power the Arduino.

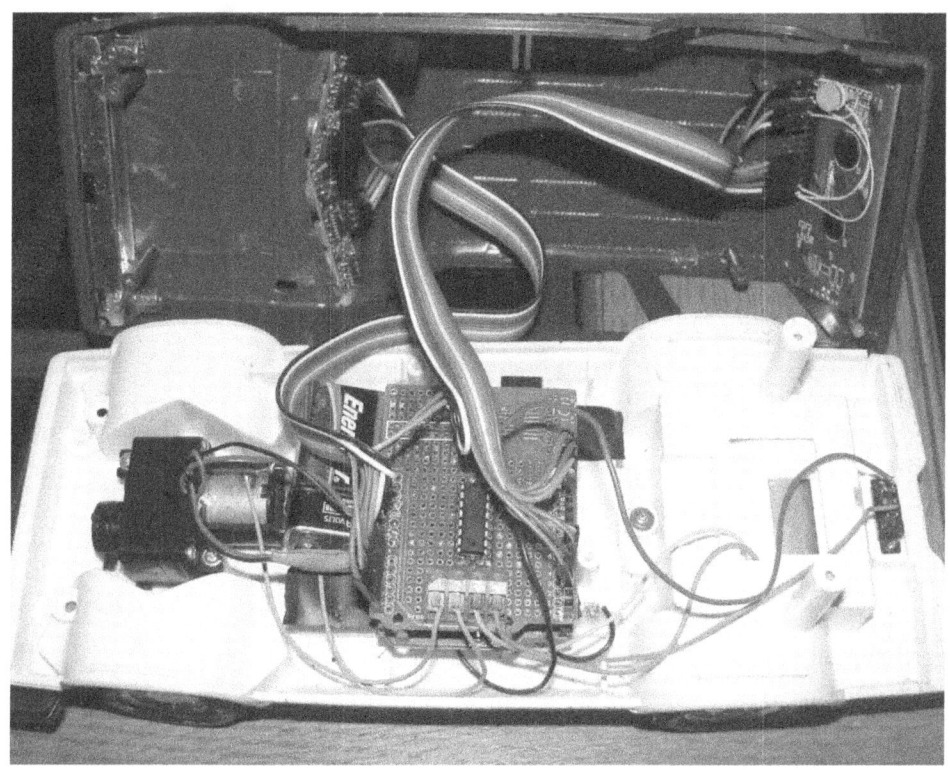

Here is the code for the two Ultrasonic detectors to display the range on the LCD and to make the car go. At a range of around 20 inches the car backs up and turns. Only one of the ultrasonic sensors was actually necessary for this mode of operation.

/******************************************
Robotics-Ultrasonic Range Car
Demonstrates use of 16x2 LCD and HC-SR04 ultrasonic range
This sketch displays two ranges on the LCD

  The LCD circuit:
* LCD RS to D 12, En to D 11
* LCD D4-D7 to D 5-2
* LCD R/W pin to ground
* Variable resistor wiper to LCD VO pin (pin 3)
  The Ultrasonic Circuit
* Echo on pin 14 (A0)
* Trig on pin 15 (A1)
* Echo2 on pin 16 (A2)
* Trig2 on pin 17 (A3)

Adapted from code written by David A. Mellis
Modified by Limor Fried, Tom Igoe, Bob Davis
*****************************************/

```
// include the library code:
#include <LiquidCrystal.h>
// initialize the library with interface pin #'s
LiquidCrystal lcd(12, 11, 5, 4, 3, 2);

#define echoPin 14 // Echo Pin
#define trigPin 15 // Trigger Pin
#define echoPin2 16 // Echo Pin 2
#define trigPin2 17 // Trigger pin 2
int durationR;
int distanceR; // used to calculate right distance
int durationL;
int distanceL; // used to calculate left distance
int motor1A =7;
int motor1B =6;
int motor2A =9;
int motor2B =10;
int mspeed =0;

void setup() {
  // set up the LCD's number of columns and rows:
  lcd.begin(16, 2);
  pinMode(trigPin, OUTPUT);
  pinMode(echoPin, INPUT);
  pinMode(trigPin2, OUTPUT);
  pinMode(echoPin2, INPUT);
  pinMode(motor1A, OUTPUT);
  pinMode(motor1B, OUTPUT);
  pinMode(motor2A, OUTPUT);
  pinMode(motor2B, OUTPUT);
}

void loop() {
// Send out ultrasonic sound
  digitalWrite(trigPin, LOW);
  delayMicroseconds(2);
  digitalWrite(trigPin, HIGH);
```

```
  delayMicroseconds(10);
  digitalWrite(trigPin, LOW);
// Wait for response back
  durationR = pulseIn(echoPin, HIGH);
//Calculate the distance in inches.
  distanceR = durationR/148;
//  delay(100);

// Second Ultrasonic ping
  digitalWrite(trigPin2, LOW);
  delayMicroseconds(2);
  digitalWrite(trigPin2, HIGH);
  delayMicroseconds(10);
  digitalWrite(trigPin2, LOW);
// Wait for response back
  durationL = pulseIn(echoPin2, HIGH);
//Calculate the distance in inches.
  distanceL = durationL/148;

// Display the results
  lcd.clear();
// Right Range
  lcd.setCursor(0,0);
  lcd.print("R Range:");
  lcd.setCursor(10, 0);
  lcd.print(distanceR);
// Left range
  lcd.setCursor(0,1);
  lcd.print("L Range:");
  lcd.setCursor(10, 1);
  lcd.print(distanceL);

// Motor control
if (distanceL > 20){
// Run straight forward
  digitalWrite(motor1A, HIGH);
  digitalWrite (motor1B, HIGH);
  digitalWrite(motor2A, HIGH);
  digitalWrite (motor2B, LOW);
}
else {
// Back up and turn
```

```
    digitalWrite(motor1A, LOW);
    digitalWrite (motor1B, HIGH);
    digitalWrite(motor2A, LOW);
    digitalWrite (motor2B, HIGH);
  }
  delay(500);
}
```

Now we will add wireless Bluetooth control to the toy car.

While deciding on what type of wireless communications devices to use for this project I looked at several different ones. I determined that the best wireless device to use is Bluetooth because you can also control it with your telephone or anything else that can send data to a Bluetooth device. Bluetooth also emulates a serial port, so data can be sent and received just like it was a direct serial or USB connection.

When the Bluetooth receiver arrived in the mail it did not came with any real instructions. All that I had to work with was these specs:

Default serial port setting: 9600 1
Pairing code: 1234
Running in slave role: Pair with BT dongle and master module
Coupled Mode: Two modules will establish communication automatically when they are powered up.
PC hosted mode: Pair the module with bluetooth dongle directly as virtual serial device.
Bluetooth protocol : Bluetooth Specification v2.0+EDR
Frequency : 2.4GHz ISM band
Modulation : GFSK(Gaussian Frequency Shift Keying)
Emission power : <=4dBm, Class 2
Sensitivity : <=-84dBm at 0.1% BER
Speed : Asynchronous: 2.1Mbps(Max) / 160 kbps,
Synchronous: 1Mbps/1Mbps
Security : Authentication and encryption
Profiles : Bluetooth serial port
CSR chip : Bluetooth v2.0
Wave band : 2.4GHz-2.8GHz, ISM Band
Protocol : Bluetooth V2.0
Power Class : (+6dbm)
Reception sensitivity: -85dBm
Voltage : 3.3 (2.7V-4.2V)

Current : Paring - 35mA, Connected - 8mA
Temperature : -40~ +105 Degrees Celsius
User defined Baud rate : 4800, 9600, 19200, 38400, 57600, 115200, 230400,460800,921600 ,1382400.
Dimension : 26.9mm*13mm*2.2mm

Setting up a Bluetooth wireless remote control is not easy to do the first time. It took me several hours to get it working the first time I tried. First, start by plugging in the Bluetooth adapter to your PC and installing the needed drivers. Usually it will automatically install the drivers over the Internet if you select that option and have an Internet connection.

Next you can power up the Bluetooth receiver module that will connect to the Arduino. Only connect the power and ground pins to the receiver for now. If you have the complete adapter it will work on either 3.3 or 5 volts, otherwise it must have 3.3 volts to operate. The receiver should start blinking a red LED to let you know that it has power.

On the PC, go to "Control panel" and select the "Bluetooth devices" icon. In the menu that comes up, select "add" and then check the box that says "It is powered up and ready to be found" and then select "next". The next picture is what you should see as the computer searches for Bluetooth devices.

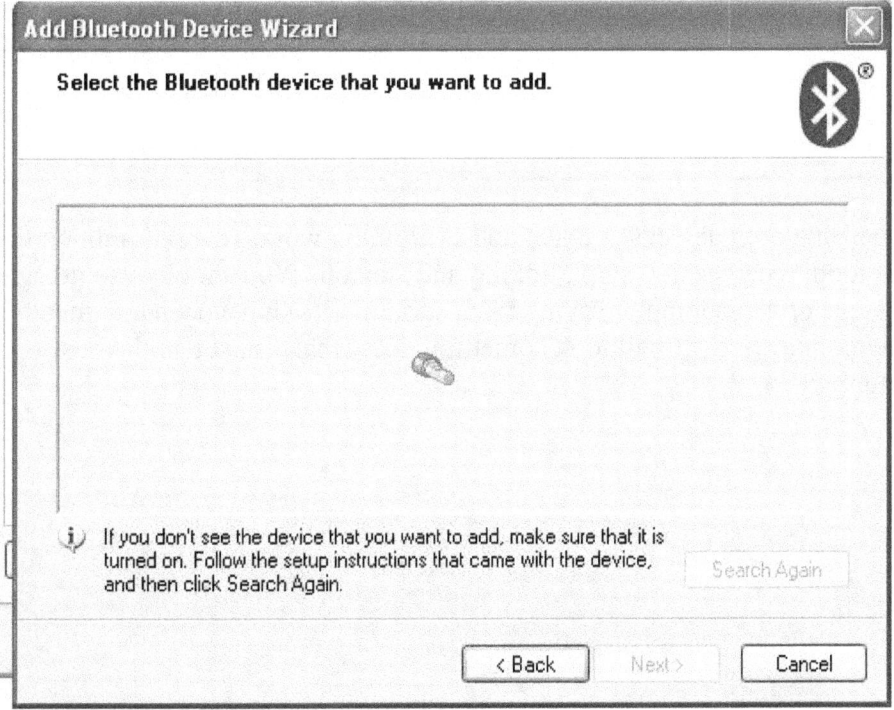

Most of the Arduino interface devices come up as "HC-06". Select it, and at the prompt, enter the default pass code of "1234". The "Add Bluetooth Device" screen looks something like this.

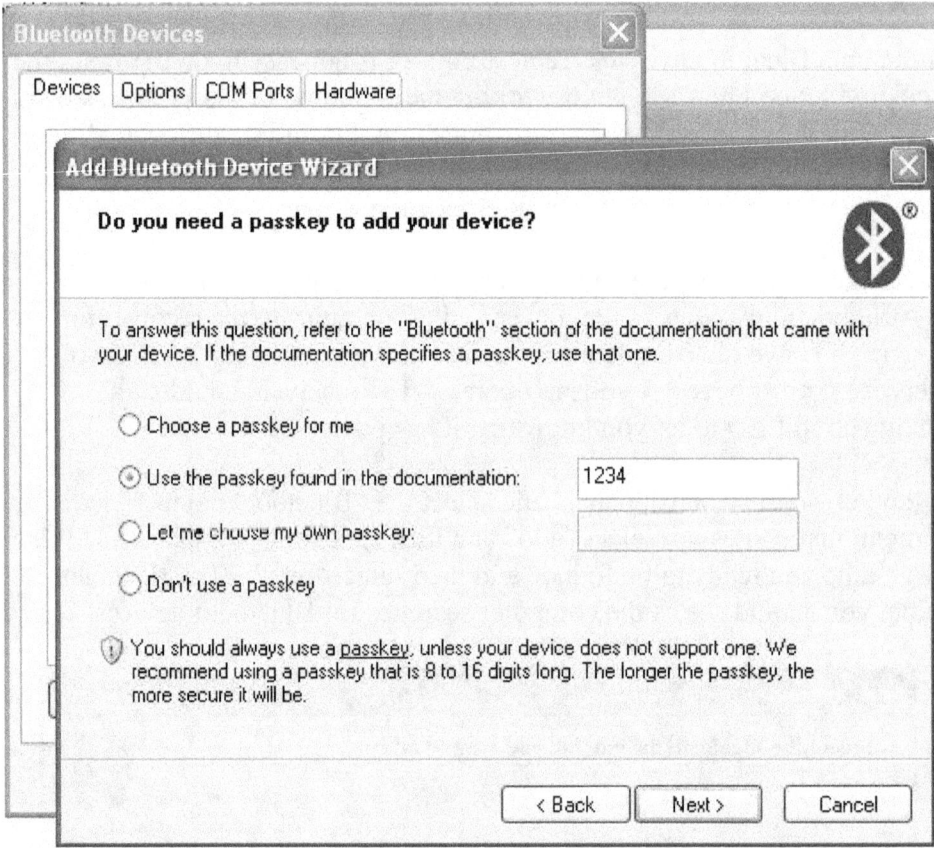

If the Bluetooth devices connect and everything works, the red light on the Bluetooth receiver will stop blinking and stay on. You should also get a message on the computer saying what serial port(s) have been assigned to the Bluetooth device. It will look something like what you see in the next picture.

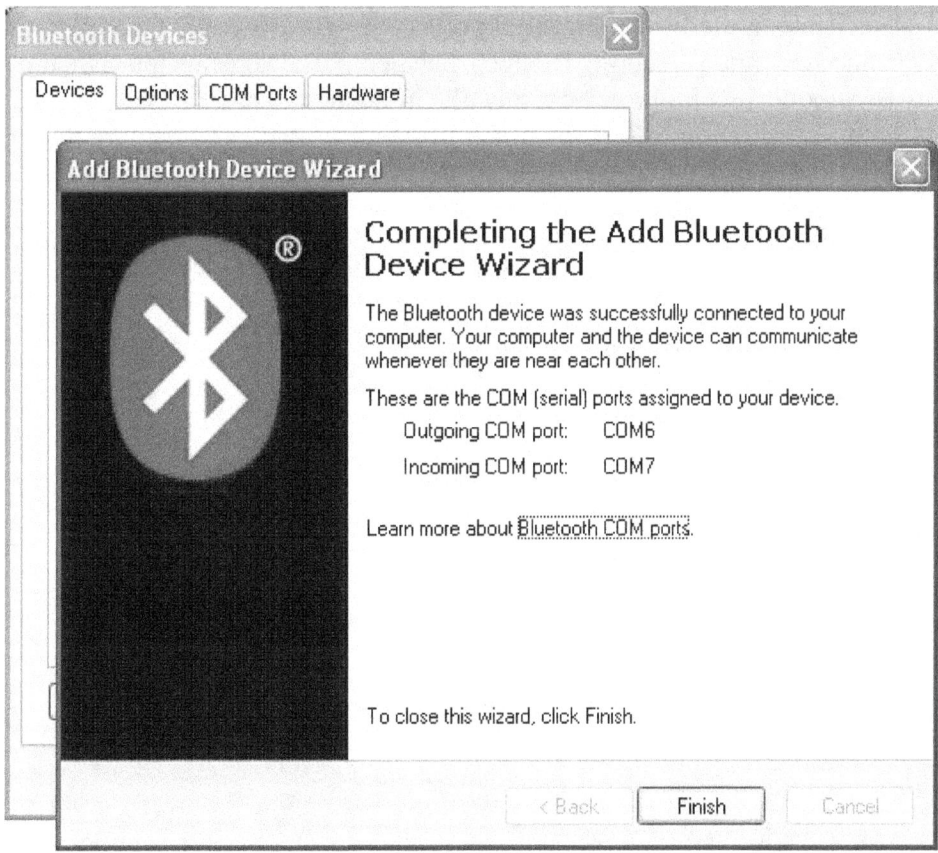

Next you need to load a sketch such as "Bluetooth test" into your Arduino. If you have a LCD connected, try the "Bluetooth to LCD" test program. Then connect the "TXD" pin of the Bluetooth receiver to the Arduino data pin D0. Note that this will block you from uploading any additional sketches until it is disconnected.

Next load a terminal program on the PC such as "Hyperterminal" and set it up for the first communications port that was assigned to your Bluetooth device, and "Serial," "9600 baud," "N," "8," and "1". Those settings should be the default settings. Save this instance of hyperterminal for future use.

At this point you should be able to type on your PC keyboard and see it on the LCD attached to the Arduino, or press "0" and "1" to see the LED on the Arduino pin 13 turn on and off. Congratulations, you have mastered connecting a PC to an Arduino via Bluetooth. I could not find any step by step instructions to do this anywhere!

You can check on the Bluetooth serial ports at any time by right clicking on "My Computer" and selecting "Device Manager". The screen should then show Bluetooth stuff in three locations like in the next picture. The Bluetooth adapter should come up under "Bluetooth Radios", "Network Adapters" and "Ports".

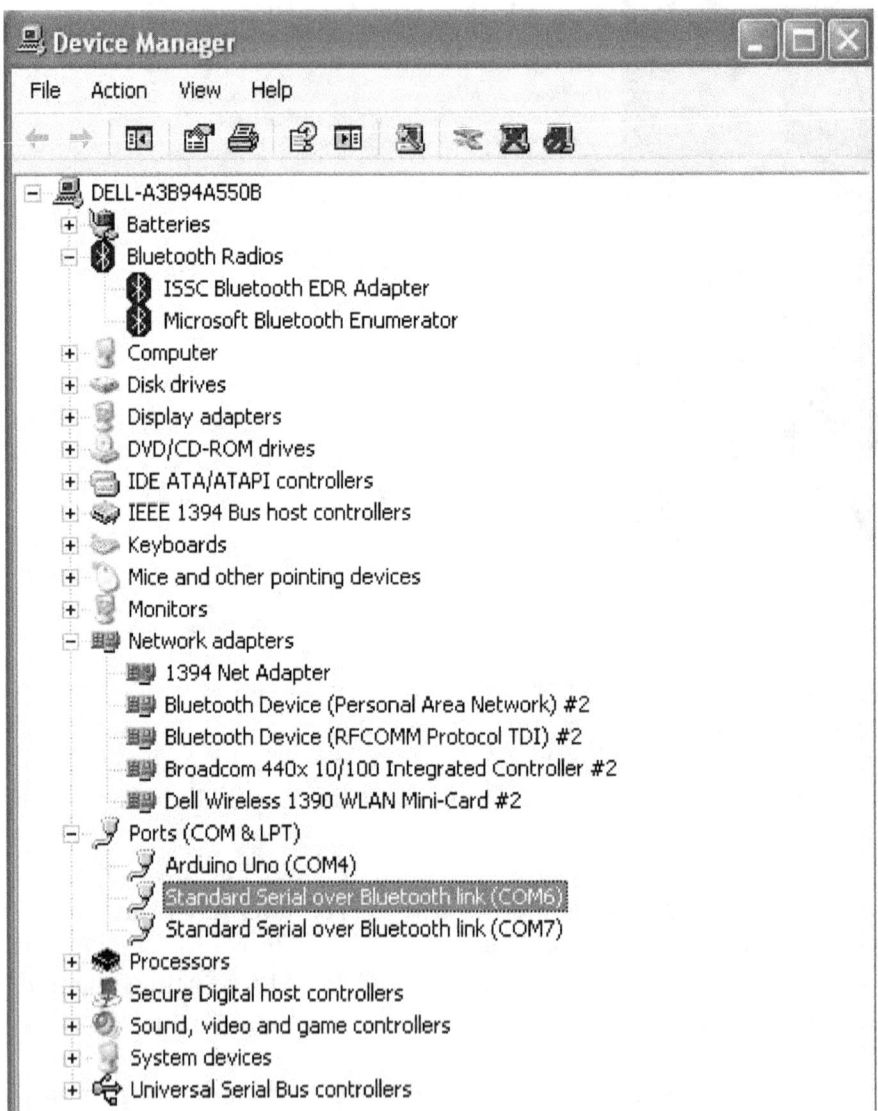

Up next is a picture showing how to connect the Bluetooth receiver to the Arduino. The Bluetooth module runs on 3.3 volts internally. **Do not connect D1 directly to the Arduino**, insert a 1K resistor, or simply do not use it. D1 is not needed to send commands from the PC to the Arduino. It is only needed to send responses back to the computer.

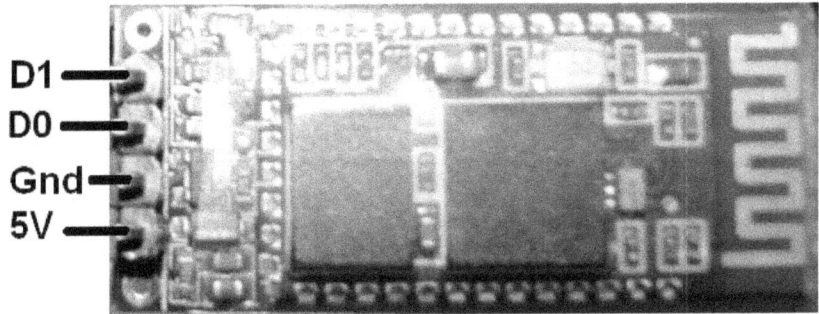

Now if you still have the dual motor controller attached to your toy car you should be able to upload this next sketch and then drive the car around using your computer keyboard.

The keyboard commands are as follows:
Forward = f
Back up = b
Right = r
Left = l
Stop = s

Here is the sketch code to make the Bluetooth controlled toy car work.

```
/*******************************************
Robotics-Serial Controlled Car
Demonstrates use of 16x2 LCD and Bluetooth serial
This sketch displays two ranges on the LCD

  The LCD circuit:
 * LCD RS to D 12, En to D 11
 * LCD D4-D7 to D 5-2
 * LCD R/W pin to ground
 * Variable resistor wiper to LCD VO pin (pin 3)

Adapted from code written by David A. Mellis
Modified by Limor Fried, Tom Igoe, Bob Davis
*******************************************/

// include the library code:
#include <LiquidCrystal.h>
// initialize the library with interface pin #'s
LiquidCrystal lcd(12, 11, 5, 4, 3, 2);
```

```
char INBYTE;
int  LED = 13; // LED on pin 13
int motor1A =7;
int motor1B =6;
int motor2A =9;
int motor2B =10;
int hpos = 0;

void setup() {
  // set up the LCD's number of columns and rows:
  lcd.begin(16, 2);
  Serial.begin(9600);
  pinMode(motor1A, OUTPUT);
  pinMode(motor1B, OUTPUT);
  pinMode(motor2A, OUTPUT);
  pinMode(motor2B, OUTPUT);
  lcd.clear();
}

void loop() {
// read next available byte
  INBYTE = Serial.read();

// Show Data
  lcd.setCursor(0,0);
  lcd.print("Data");
if (INBYTE != ' ') {
  lcd.setCursor(6+hpos, 0);
  hpos=hpos++;
  if (hpos > 10) hpos=0;
  lcd.print(INBYTE);
}
// Motor control
if (INBYTE == 'f'){
// Run straight forward
  digitalWrite(motor1A, HIGH);
  digitalWrite (motor1B, HIGH);
  digitalWrite(motor2A, HIGH);
  digitalWrite (motor2B, LOW);
}
if (INBYTE == 'b'){
// Back up and turn
```

```
    digitalWrite(motor1A, HIGH);
    digitalWrite (motor1B, HIGH);
    digitalWrite(motor2A, LOW);
    digitalWrite (motor2B, HIGH);
 }
 if (INBYTE == 's'){
 // Back up and turn
    digitalWrite(motor1A, HIGH);
    digitalWrite (motor1B, HIGH);
    digitalWrite(motor2A, HIGH);
    digitalWrite (motor2B, HIGH);
 }
 if (INBYTE == 'r'){
 // Back up and turn
    digitalWrite(motor1A, HIGH);
    digitalWrite (motor1B, LOW);
    digitalWrite(motor2A, HIGH);
    digitalWrite (motor2B, LOW);
 }
 if (INBYTE == 'l'){
 // Back up and turn
    digitalWrite(motor1A, LOW);
    digitalWrite (motor1B, HIGH);
    digitalWrite(motor2A, HIGH);
    digitalWrite (motor2B, LOW);
 }
    delay(500);
 }
```

Once you have the Bluetooth controlled car up and running, you might want to add a wireless CCTV (Closed Circuit TV) camera. With that addition you could drive the car through tight spaces and see what is in there on a TV set. You can get a complete wireless camera transmitter and receiver kit on eBay for around $25. What you will need is the camera, power adapter, cables, and a receiver. Up next is a picture of a typical wireless CCTV camera, with six IR night vision LED's. It takes a fresh new nine volt battery to power the camera.

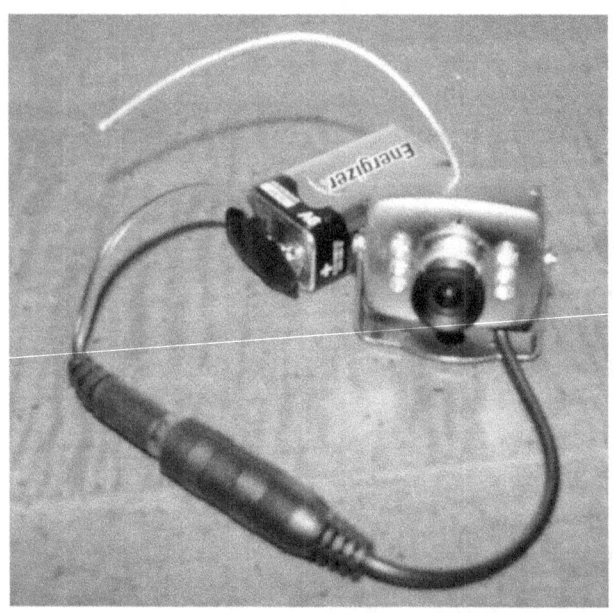

Here is a picture of the Bluetooth controlled car with the wireless camera mounted on the roof. You could even add a servo motor so you would be able to rotate the camera right and left by up to 90 degrees.

This next picture is of the wireless receiver. It uses a 12 volt DC adapter for power. There is a silver frequency knob that has to be tuned for the best picture quality. The yellow and white RCA cables go to your TV set. You can also connect the RCA cables to a USB video adapter and then you can watch what is happening on a computer screen.

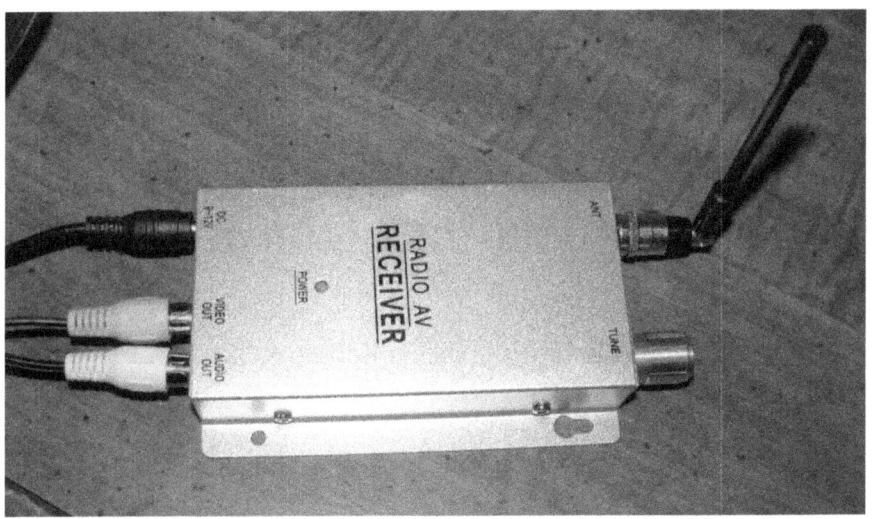

The resulting picture quality is not that great. The cheap camera lacks good low level light sensitivity. Here is a typical screen shot from a TV attached to the receiver using room lighting. You can see the lack of picture quality when just using typical room lighting. Maybe you could add headlights to the car to give some better lighting?

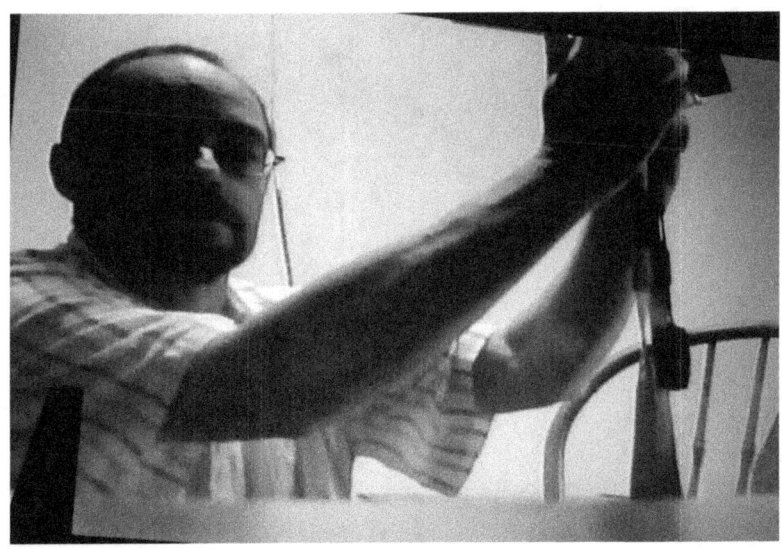

# Chapter 7

# Arduino Controlled Roomba Robot

This chapter will show you how to make a Roomba Vacuum Cleaner into an Arduino powered Robot.

The drive mechanism of a Roomba vacuum cleaner makes a good start or base for building a robot. The motors are very powerful and they are very rugged. It takes a bit of work to take apart but once it is disassembled there are lots of wires for you to figure out where they go and what they do.

Here is a picture of some of the parts that were removed from the Roomba.

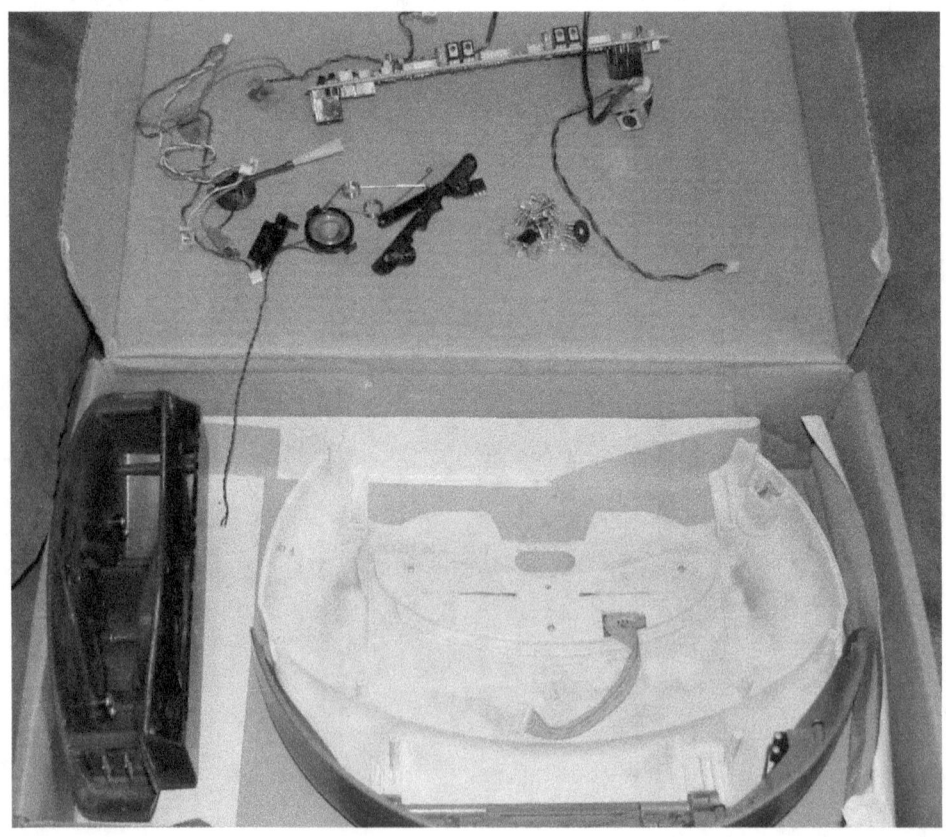

What I used was an iRobot Roomba model number 4210. This is a picture of the stripped Roomba. Do you see what looks like a dinosaur foot in there? It just needs a center toe. One of my long term goals is to make a life sized robotic Raptor dinosaur and this Roomba might well become one of its feet!

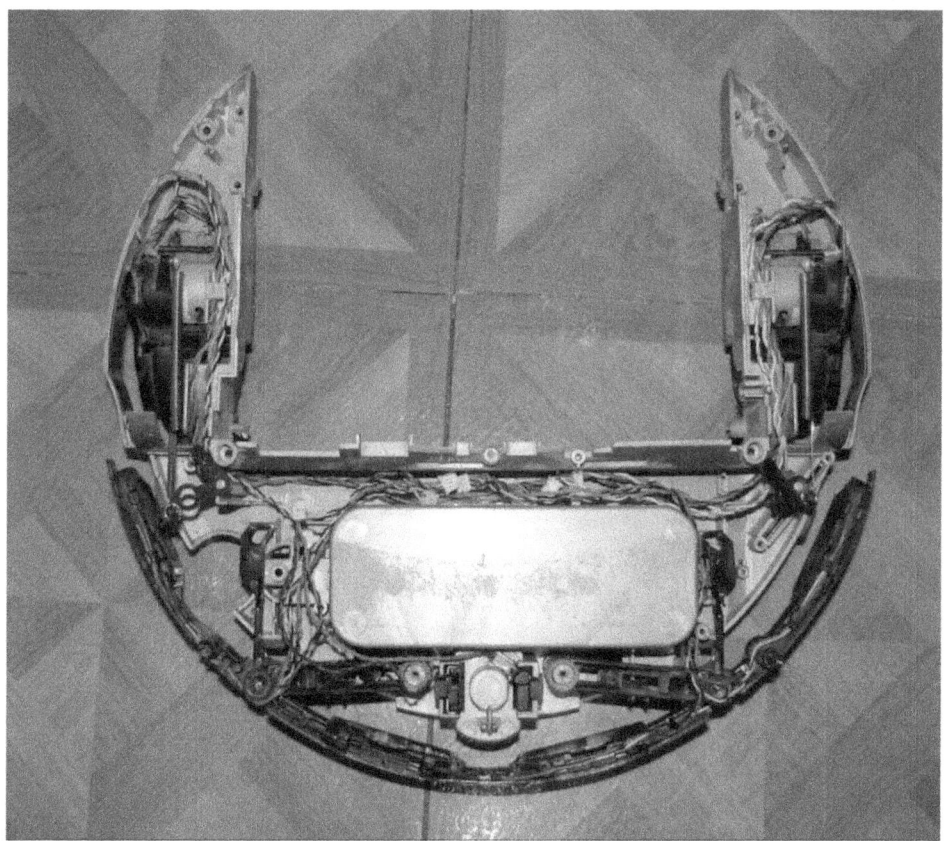

This is a color code listing of the wires that we will be using.

Left side motor:
--------------------
Red and Orange – Power to the motor
Blue/Brown/Gray/Black – Wheel movement sensor
Yellow and black – Wheel up/down switch

Right side motor:
--------------------
Red and Orange – Power to the motor
Blue/Brown/Gray/Black – Wheel movement sensor
White/Gray – Wheel up/down switch

Front Wheel sensors:
--------------------------
Green/Green – wheel up/down switch
Purple/Purple - Unknown

To get this up and running you can use a home made L293 IC based shield. The L293 gets its inputs from the Arduino D6, D7, D9, and D10 pins. Connect jumpers from the Roomba's jacks where the red and orange wires are connected to the outputs of the motor controller. Reverse the polarity of the wires if the motor turns in the wrong direction. You can use four or six "AA" batteries in a battery holder as the power source.

Here is a sketch to make it run under serial control.

```
/*******************************************
Robotics-Remote control 2 wheel robot
Demonstrates the use of serial control
Written by Bob Davis

The serial commands:
 f=forward
 b=back up
 r=right
 l=left
 s=stop
*******************************************/

char INBYTE;
int motor1A =6;
int motor1B =7;
int motor2A =9;
int motor2B =10;

void setup() {
  Serial.begin(9600);
  pinMode(motor1A, OUTPUT);
  pinMode(motor1B, OUTPUT);
  pinMode(motor2A, OUTPUT);
  pinMode(motor2B, OUTPUT);
}
```

```
void loop() {
// read next available byte
  INBYTE = Serial.read();
// Motor control
if (INBYTE == 'f'){
// Run straight forward
  digitalWrite(motor1A, HIGH);
  digitalWrite (motor1B, LOW);
  digitalWrite(motor2A, HIGH);
  digitalWrite (motor2B, LOW);
}
if (INBYTE == 'b'){
// Back up
  digitalWrite(motor1A, LOW);
  digitalWrite (motor1B, HIGH);
  digitalWrite(motor2A, LOW);
  digitalWrite (motor2B, HIGH);
}
if (INBYTE == 's'){
// stop all motors
  digitalWrite(motor1A, HIGH);
  digitalWrite (motor1B, HIGH);
  digitalWrite(motor2A, HIGH);
  digitalWrite (motor2B, HIGH);
}
if (INBYTE == 'r'){
// Back up and turn
  digitalWrite(motor1A, HIGH);
  digitalWrite (motor1B, LOW);
  digitalWrite(motor2A, LOW);
  digitalWrite (motor2B, HIGH);
}
if (INBYTE == 'l'){
// Back up and turn
  digitalWrite(motor1A, LOW);
  digitalWrite (motor1B, HIGH);
  digitalWrite(motor2A, HIGH);
  digitalWrite (motor2B, LOW);
}
  delay(500);
}
```

I added an eight inch long wooden board across the middle or the Roomba to mount things to. I also added a big 12 volt battery pack weighing about eight pounds, an Arduino, and a home made motor controller. Sticking up in the air above the Arduino on the right side is a push button switch used for troubleshooting the code. Then I mounted an ultrasonic sensor in the middle and took this picture of the Arduino powered Roomba robot.

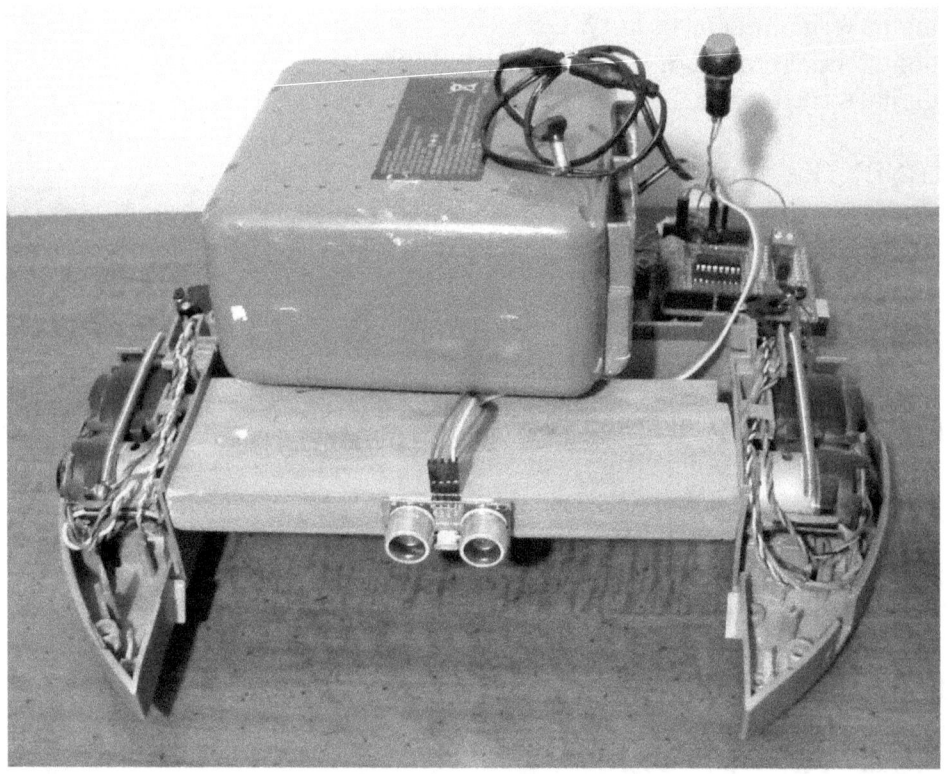

This program demonstrates the Roomba running on its own with the ultrasonic distance sensing. It detects and avoids objects at about 12 inches distance. I used the HC-SR04 ultrasonic sensor that was covered in an earlier chapter. It was mounted in the center of the front of the Roomba.

/*******************************************
Robotics-Roomba Roamer controls 2 wheel robot
Demonstrates the Roomba and ultrasonic distance sensing

The sequence of commands:
 Check for collision?
 if not: forward
 else: back up,

       turn right, stop, check distance
       turn left, stop, check distance
       if right distance was better then turn right

echo is on pin 16 (A2)
trigger is on pin 17 (A3)

Written by Bob Davis January 2014
*****************************************/

int motor1A =6;
int motor1B =7;
int motor2A =9;
int motor2B =10;
#define echoPin 16
#define trigPin 17
int distance=10;
int distanceR=10;
int distanceL=10;

void setup() {
// Serial.begin(9600);
  pinMode(motor1A, OUTPUT);
  pinMode(motor1B, OUTPUT);
  pinMode(motor2A, OUTPUT);
  pinMode(motor2B, OUTPUT);
  pinMode(trigPin, OUTPUT);
  pinMode(echoPin, INPUT);
}

void loop() {
// Wait for the switch to start demo
// Used for debugging the sketch
// while (analogRead(A0) > 10){}

// Stop all motors
  digitalWrite(motor1A, HIGH);
  digitalWrite (motor1B, HIGH);
  digitalWrite(motor2A, HIGH);
  digitalWrite (motor2B, HIGH);
// send out ultrasonic ping
  digitalWrite(trigPin, LOW);

```
  delayMicroseconds(2);
  digitalWrite(trigPin, HIGH);
  delayMicroseconds(10);
  digitalWrite(trigPin, LOW);
// Wait for response back
  distance=pulseIn(echoPin, HIGH);

  if (distance > 2000){
// Run straight forward
    digitalWrite(motor1A, HIGH);
    digitalWrite (motor1B, LOW);
    digitalWrite(motor2A, HIGH);
    digitalWrite (motor2B, LOW);
    delay(1000);
    }
  else {
// Back up a little
    digitalWrite(motor1A, LOW);
    digitalWrite (motor1B, HIGH);
    digitalWrite(motor2A, LOW);
    digitalWrite (motor2B, HIGH);
    delay(200);
// Turn right
    digitalWrite(motor1A, HIGH);
    digitalWrite (motor1B, LOW);
    digitalWrite(motor2A, LOW);
    digitalWrite (motor2B, HIGH);
    delay(500);
// Stop all motors
    digitalWrite(motor1A, HIGH);
    digitalWrite (motor1B, HIGH);
    digitalWrite(motor2A, HIGH);
    digitalWrite (motor2B, HIGH);
// send out ultrasonic ping to check distance
    digitalWrite(trigPin, LOW);
    delayMicroseconds(2);
    digitalWrite(trigPin, HIGH);
    delayMicroseconds(10);
    digitalWrite(trigPin, LOW);
// Wait for response back
    distanceR=pulseIn(echoPin, HIGH);
// Turn left
```

```
    digitalWrite(motor1A, LOW);
    digitalWrite (motor1B, HIGH);
    digitalWrite(motor2A, HIGH);
    digitalWrite (motor2B, LOW);
    delay(1000);
// Stop all motors
    digitalWrite(motor1A, HIGH);
    digitalWrite (motor1B, HIGH);
    digitalWrite(motor2A, HIGH);
    digitalWrite (motor2B, HIGH);
// send out ultrasonic ping to check distance
    digitalWrite(trigPin, LOW);
    delayMicroseconds(2);
    digitalWrite(trigPin, HIGH);
    delayMicroseconds(10);
    digitalWrite(trigPin, LOW);
// Wait for response back
    distanceL=pulseIn(echoPin, HIGH);
    if (distanceL < distanceR){
// Go back to the right
    digitalWrite(motor1A, HIGH);
    digitalWrite (motor1B, LOW);
    digitalWrite(motor2A, LOW);
    digitalWrite (motor2B, HIGH);
    delay(1000);
    }
  }
}
```

Another method of controlling your robot is to use an IR remote control much like the one you use for your TV. I purchased a common IR remote with an IR receiver that I found on eBay and then figured out what the codes are that it sends. Then I modified the Roomba sketch so it will work with this IR remote control. At first the Roomba continued to execute the last command until it received another command, but that led to it running into things. So now the sketch will stop after it does each command and wait for the next command before it proceeds.

IR remote control codes can be quite complex. They use frequencies from 36 KHz to 40 KHz and send a wide variety of number of pulses. Some use positive pulses and some use negative pulses. Most IR remotes use missing pulses to either signify data of a one or a zero.

With the remote control that we are using it sends out about 32 pulses before it sends out the data. Those pulses show up as several F's ahead of the code when you are looking at the raw data. Those F's are hexadecimal code for four 1's in a row. Here is what a typical IR command might look like.

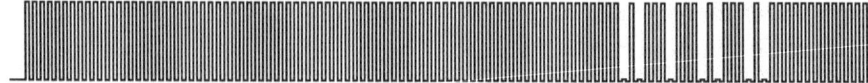

Fortunately for us there is an IR Remote control library available for the Arduino. The remote control library will look at the raw IR data and determine what the "standard" is being used. Then it can remove the unneeded information and just give you the data that you need to interpret the command that is being sent. The library is called "IRremote.zip" and it needs to be downloaded and unzipped into your Arduino library to be able to use it.

Here is a picture of the IR transmitter and receiver that were used to control the Roomba. The receiver pin connections are negative, positive, and signal from left to right when looking at the back of the receiver. They are not all clearly marked on the circuit board. The only markings are "-" for negative and "S" for signal.

Here is an IR remote control demonstration program.
/*******************************************
Robotics-Roomba IR Remote control of 2 wheel robot
Demonstrates the Roomba under IR wireless
Written by Bob Davis January 2014
*******************************************/

```
// IR setup
#include <IRremote.h>
int RECV_PIN = 11;
IRrecv irrecv(RECV_PIN);
decode_results results;

//char INBYTE;
int motor1A =6;
int motor1B =7;
int motor2A =9;
int motor2B =10;

void setup() {
  Serial.begin(9600);
  irrecv.enableIRIn(); // Start the receiver
  pinMode(motor1A, OUTPUT);
  pinMode(motor1B, OUTPUT);
  pinMode(motor2A, OUTPUT);
  pinMode(motor2B, OUTPUT);
}

void loop() {
  if (irrecv.decode(&results)) {
    if (results.value==16738455) Serial.println("1 key");
    if (results.value==16750695) Serial.println("2 key");
    if (results.value==16756815) Serial.println("3 key");
    if (results.value==16724175) Serial.println("4 key");
    if (results.value==16718055) Serial.println("5 key");
    if (results.value==16743045) Serial.println("6 key");
    if (results.value==16716015) Serial.println("7 key");
    if (results.value==16726215) Serial.println("8 key");
    if (results.value==16734885) Serial.println("9 key");
    if (results.value==16730805) Serial.println("0 key");
    if (results.value==16728765) Serial.println("star key");
```

```
if (results.value==16732845) Serial.println("pound key");
if (results.value==16712445) Serial.println("OK key");
if (results.value==16736925) {
  Serial.println("forward key");
  // Go forward
  digitalWrite(motor1A, HIGH);
  digitalWrite (motor1B, LOW);
  digitalWrite(motor2A, HIGH);
  digitalWrite (motor2B, LOW);
  delay(2000);
}
if (results.value==16761405) {
  Serial.println("right key");
  // Turn right
  digitalWrite(motor1A, HIGH);
  digitalWrite (motor1B, LOW);
  digitalWrite(motor2A, LOW);
  digitalWrite (motor2B, HIGH);
  delay(1000);
}
if (results.value==16754775) {
  Serial.println("back key");
  // Go Backwards
  digitalWrite(motor1A, LOW);
  digitalWrite (motor1B, HIGH);
  digitalWrite(motor2A, LOW);
  digitalWrite (motor2B, HIGH);
  delay(2000);
}
if (results.value==16720605) {
  Serial.println("left key");
  // Turn left
  digitalWrite(motor1A, LOW);
  digitalWrite (motor1B, HIGH);
  digitalWrite(motor2A, HIGH);
  digitalWrite (motor2B, LOW);
  delay(1000);
}
// Stop all motors
digitalWrite(motor1A, HIGH);
digitalWrite (motor1B, HIGH);
digitalWrite(motor2A, HIGH);
```

digitalWrite (motor2B, HIGH);

// Next two lines are for troubleshooting so you can see the codes
//   Serial.println(results.value);
//   dump(&results);
  irrecv.resume(); // Receive the next value
  }
}

The L293 motor driver IC is limited to one half an amp of maximum power. For more power the L293 driver can be replaced with a better L298 IC. The L298 is a 2.5 amp per motor driver IC. Unfortunately the L298 comes with pins that do not easily fit into a breadboard. On the other hand it is available for just a few dollars already mounted on a small circuit board that makes installing it a breeze. The circuit board even has a five volt voltage regulator on it as well as diode protection of all the output pins.

Here is a picture of what that L298 circuit board looks like.

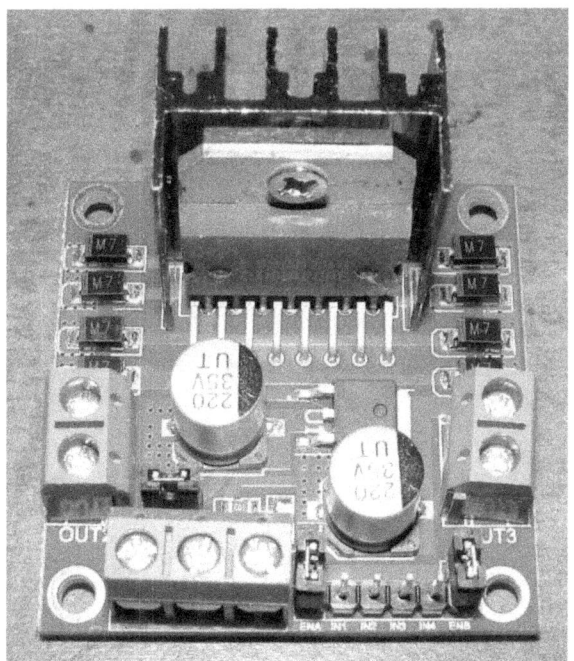

The pin definitions across the front of the board are 9-12 volts in, Ground, 5 volts out to the Arduino (optional). Then there are four data pins that go to D6 and D7 or D9 and D10 pins on the Arduino for the Roomba sketch to run properly. The motors are connected to the blue screw terminals on each side

65

of the circuit board. There are jumpers on each side of the data input pins that can be removed to allow access to the motor enable pins if they are needed.

When testing out this board, the Roomba's "right" and "left" turn commands caused it to overshoot. It turned about 110 degrees instead of 90 degrees like it used to do. This overshoot is clear indication that there is now more power going to the motors.

# Chapter 8

# Arduino Controlled Homemade Car

You can also build your own robotic car to control. There are kits for one on eBay or you can design and build your own. This chapter will cover my design that is based loosely on the designs found on the Internet. I used pictures found on the Internet then enlarged them to the correct size and designed this car around that. The small geared motors are best bought but everything else in the car body can be made.

You will need four motor mounting brackets. I made mine out of some 1.25 by 1.25 inch angled aluminum cut into pieces about 1.25 inches long. The sizes of the holes are 1/8 inch for 4-40 screws, 3/16 inch and 1/2 inch. If the 1/2 inch hole is a problem you can just cut out that corner of the bracket. The .30 measurement is 5/16 of an inch and the .69 measurement is 11/16 of an inch. If I had it to do it over again I would just buy the brackets.

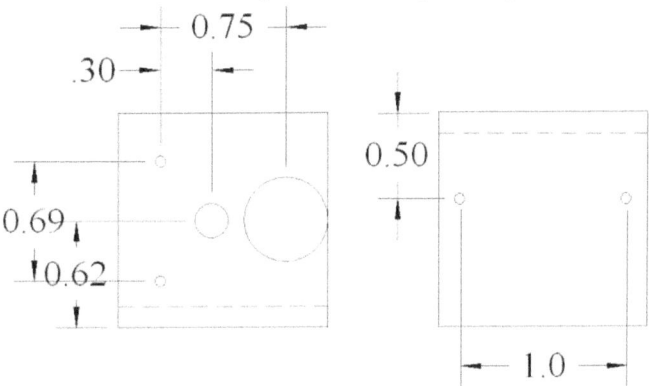

The body of the car is made out of a piece of Plexiglas about 10.5 by 6.5 inches in size. The pattern for the car can be enlarged, cut out, and then traced on the Plexiglas. I used a hand held jig saw to cut it out but a better saw would do a better job. The Plexiglas piece I used was not 6.5 inches wide so the resulting car is a little narrower.

The next page has the mechanical drawing of the body of the homemade car.

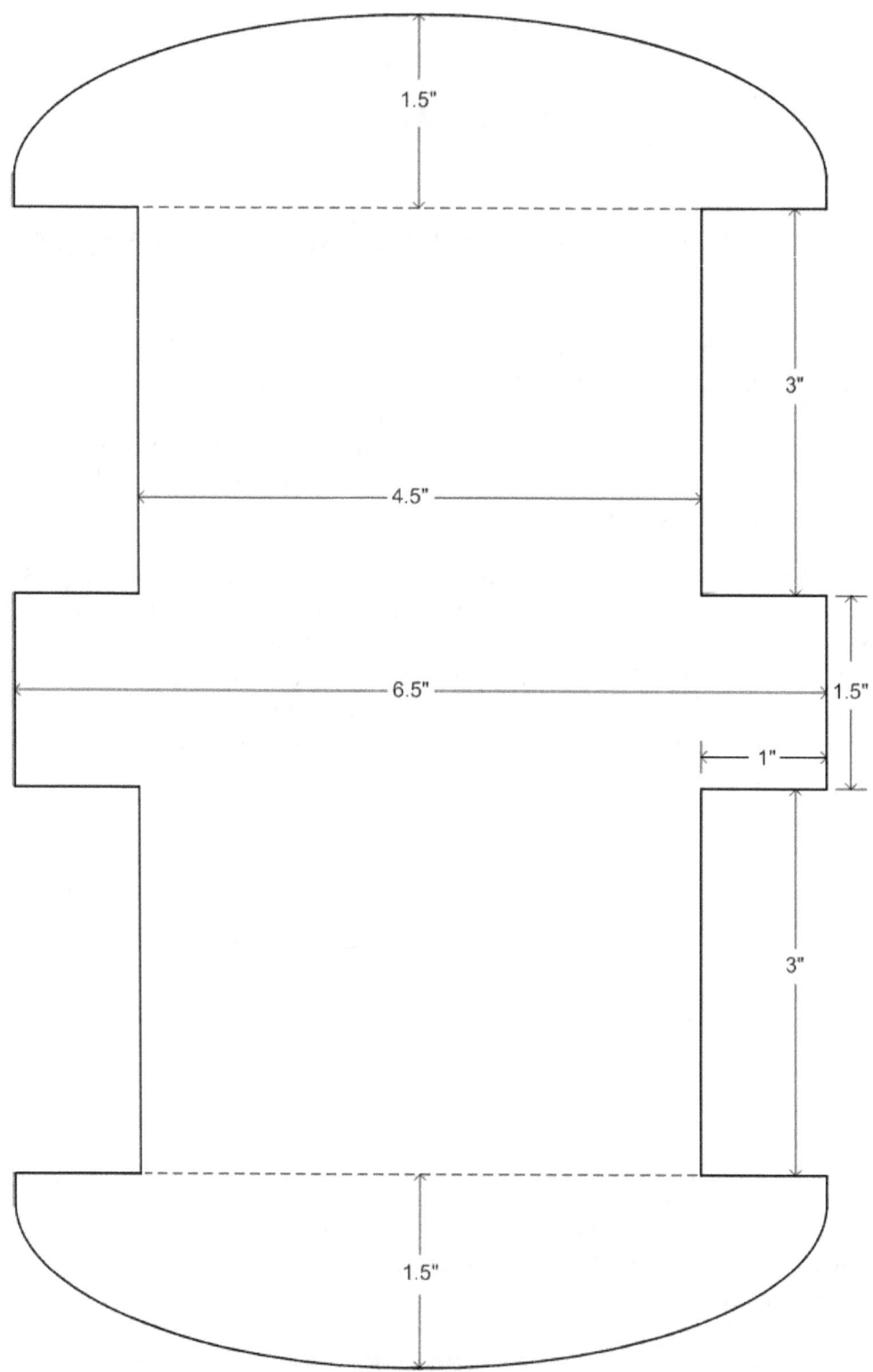

Here is a picture of the bottom view of the car body. I was hoping there would be room for batteries down there but the space was too limited.

This picture shows the top view of the homemade car before the electronics were added.

Next up is a picture of the completed car. It uses the Adafruit motor controller that is discussed in detail in another chapter. We need this controller to control all four of the motors. The ultrasonic distance sensor was installed for looks but was not implemented in the software. The motors are wired up clockwise with motor one being the front right motor.

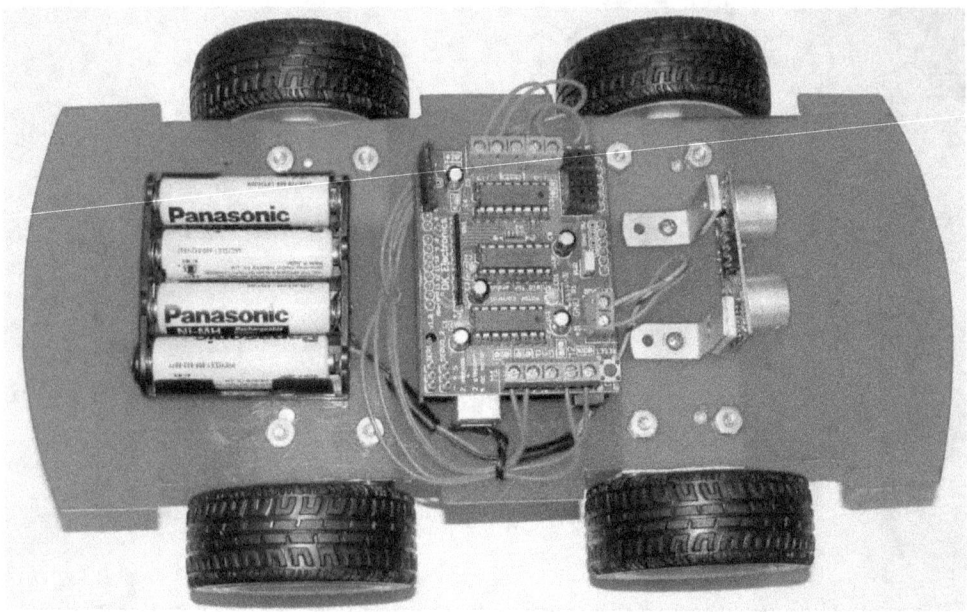

Here is the code to make the robotic car run.

```
/*******************************************
Robotics-Remote control 4 wheel robot
Demonstrates the use of serial control

The serial commands are:
 f=forward
 b=back up
 r=right
 l=left
 s=stop

Written by Bob Davis
*******************************************/
#include <AFMotor.h>
// motors are connected in a circle like a clock
// motor 1 is right front
// motor 2 is right back
// motor 3 is left back
// motor 4 is left front
```

```
AF_DCMotor motor1(1, MOTOR12_64KHZ);
AF_DCMotor motor2(2, MOTOR12_64KHZ);
AF_DCMotor motor3(3, MOTOR12_64KHZ);
AF_DCMotor motor4(4, MOTOR12_64KHZ);

void setup() {
 Serial.begin(9600);
 motor1.setSpeed(255);
 motor2.setSpeed(255);
 motor3.setSpeed(255);
 motor4.setSpeed(255);
}

void loop() {
// read next available byte
 int INBYTE = Serial.read();
// Motor control
 if (INBYTE == 'f'){
// Run straight forward
 motor1.run(FORWARD);
 motor2.run(FORWARD);
 motor3.run(FORWARD);
 motor4.run(FORWARD);
 }
 if (INBYTE == 'b'){
// Back up
 motor1.run(BACKWARD);
 motor2.run(BACKWARD);
 motor3.run(BACKWARD);
 motor4.run(BACKWARD);
 }
 if (INBYTE == 's'){
// stop all motors
 motor1.run(RELEASE);
 motor2.run(RELEASE);
 motor3.run(RELEASE);
 motor4.run(RELEASE);
 }
 if (INBYTE == 'l'){
// turn left
 motor1.run(FORWARD);
 motor2.run(FORWARD);
```

```
    motor3.run(BACKWARD);
    motor4.run(BACKWARD);
    delay(300);
    motor1.run(RELEASE);
    motor2.run(RELEASE);
    motor3.run(RELEASE);
    motor4.run(RELEASE);
   }
   if (INBYTE == 'r'){
// turn right
    motor1.run(BACKWARD);
    motor2.run(BACKWARD);
    motor3.run(FORWARD);
    motor4.run(FORWARD);
    delay(300);
    motor1.run(RELEASE);
    motor2.run(RELEASE);
    motor3.run(RELEASE);
    motor4.run(RELEASE);
   }
}
```

# Chapter 9

# Arduino Controlled RoboRaptor

In this chapter we will be rebuilding a RoboRaptor with an Arduino. This monster can be quite intimidating. I looked inside the RoboRaptor and it looked overwhelming with all of the wires that are in there. As I was testing I discovered that some of the wires just go to switches to change his mood and direction.

Here is a picture of the Robo-Raptor brochure:

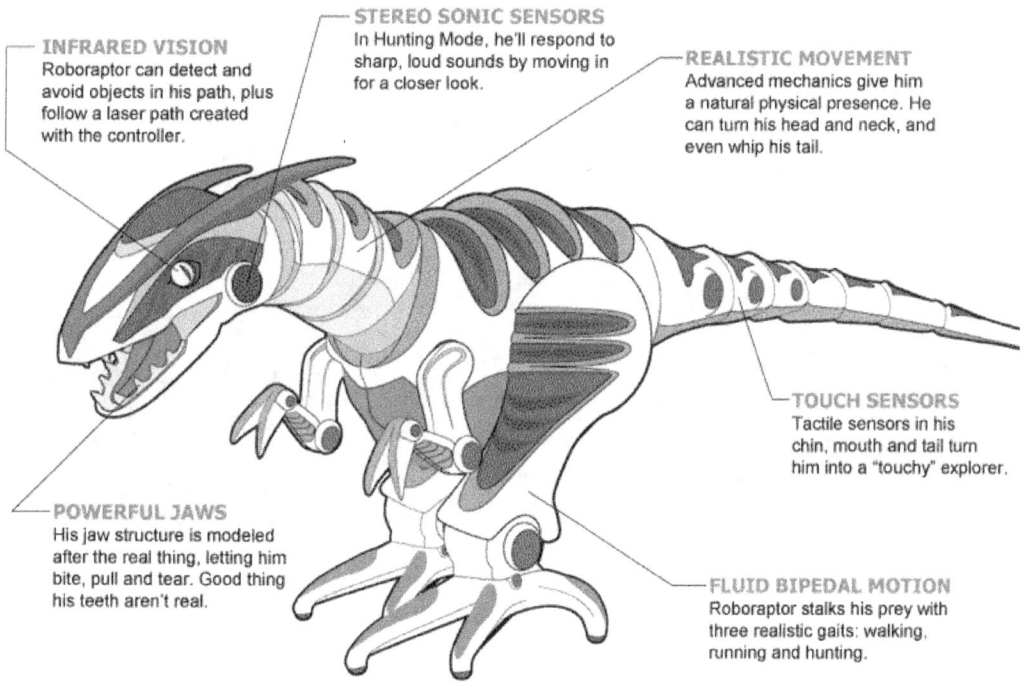

To make the RoboRaptor come to life there are the following devices:

Five - Reversible DC motors
Six - Contact switches in the tail and in the head
Seven - Motor position sensing switches.
Two - Red LED's in its nose
Two – Microphones for its ears
Two - Infrared transmitters and receivers on top its head
One – Speaker located behind its mouth

Here is what the RoboRaptor looks like with the leg and top covers removed. You have to remove the outer part of the legs, then the inner part of the legs to get inside of it. As you can see there are a lot of wires in there! I also disassembled the head and figured out what all of those wires going to the head are there for. Then I used a nine volt battery and jumper wires to verify exactly what each of the motors does when powered up.

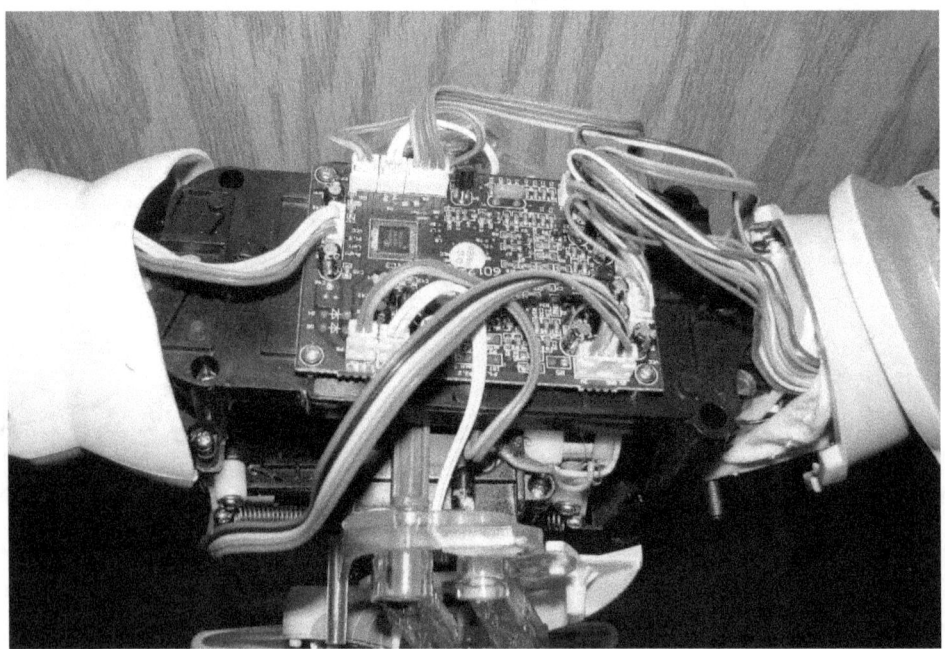

Here is a location guide for the ten connectors on the old controller board:

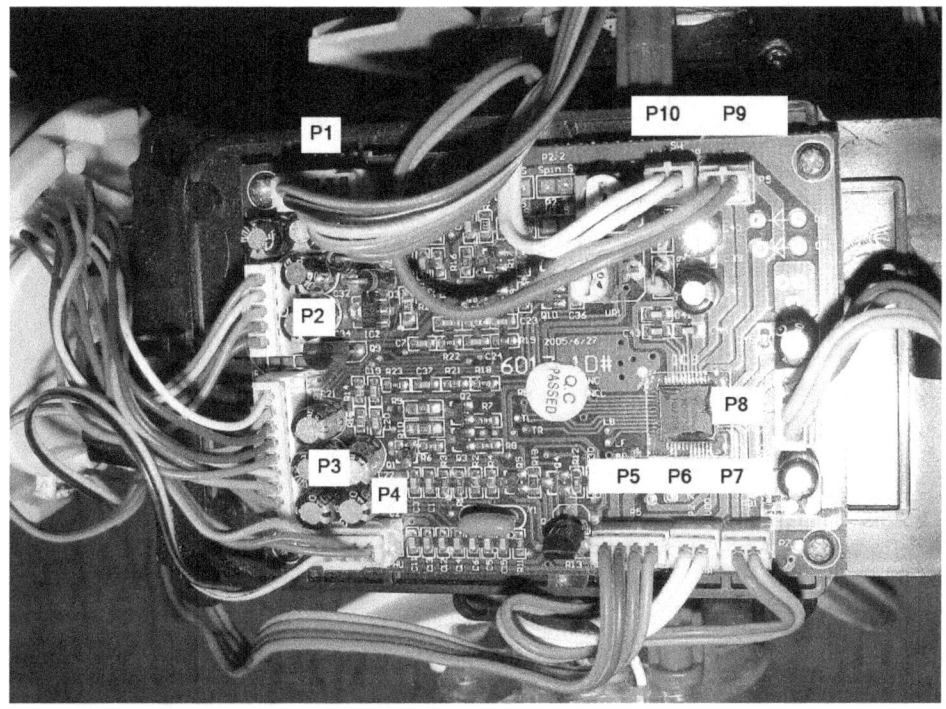

Here is a breakdown of the RoboRaptor wiring connectors and color codes:

Battery and power switch area, five pins P1:
Brown – Six volts (Need to modify for nine volts)
Black – Ground
Yellow – Black when off?
Green – Red when off?
Red – Three volts

Head area where there are two connectors.
Head connector one - five pins P2:
Green – Ground (To IR Receivers)
Yellow – Vin Power (One side of speaker, LED's)
Orange – Inside of the mouth switch (Other side goes to Vdd)
Red – Under the mouth switch (Other side goes to Vdd)
Brown – Speaker (Other side goes to Vin)

Head connector two - nine pins P3:
White – Vdd Power (One side of mouth switches)
Grey – Nose LED (other side to Vin)

Purple – Nose LED (other side to Vin)
Blue – Microphone (other side to ground)
Green - Microphone (other side to ground)
Yellow – IRV Receiver
Orange – IRQ Receiver
Red – IRV Receiver
Brown – IRQ Receiver

Additional five pins on left side P4:
Black – Ground
Yellow – Upper neck switch
Orange – Lower neck left rotation switch
Red – Lower neck center rotation switch
Brown – Lower neck right rotation switch

Additional four pins on left side P5:
Red – Rotate Neck in circles, open and close mouth Motor
Purple – Rotate Neck in circles, open and close mouth Motor
Red – Shake head up and down motor
Green – Shake head up and down motor

Left leg dual two pins P6, P7:
Red – Leg walking motor
Green – Leg walking motor
Dual white – Motor centered position sensor

Tail area four pins P8:
Red – Swing tail and neck motor
Green – Swing tail and neck motor
Dual Yellow – Tail contact switches, there are two in parallel

Right leg dual two pins P9, P10:
Red – Leg walking motor
Green – Leg walking motor
Dual white – Motor centered position sensor

Up next is what his head looks like when it is disassembled. The unplugged connector is the speaker. The yellow/black pairs are the microphones. The blue wires going to a gray switch on the far right are for the inside of the mouth switch to detect that it is biting something.

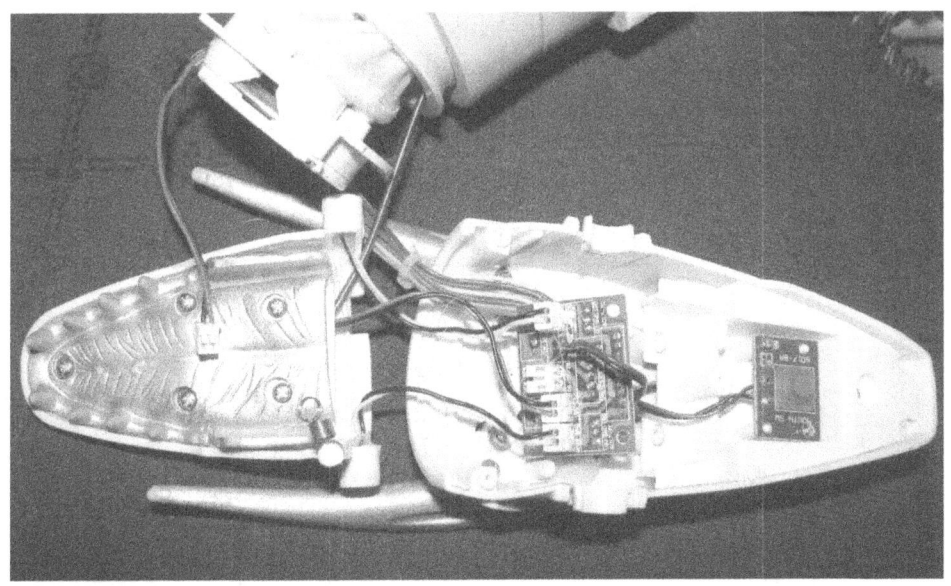

Controlling all five motors will be an issue. Our home made motor controller can only control up to two motors. So for this project we will need to step up to a four motor controller Arduino Shield. This shield is commonly available on eBay. It uses two of the L293 motor controller IC's to control up to four motors. For the fifth motor we would have to use a power transistor and only have the motor be on or off. The motor that wiggles the neck and opens the mouth would work well with that limitation.

The schematic of the four motor controller circuit board is normally very large in size and appears quite complicated. I have broken it down into three smaller partial schematics in order to make it easier to understand. The first part of the schematic is the two L293D motor drivers. The pin description on top of the pins is for the first L293 IC and the pin designation below the pins is for the second L293 IC.

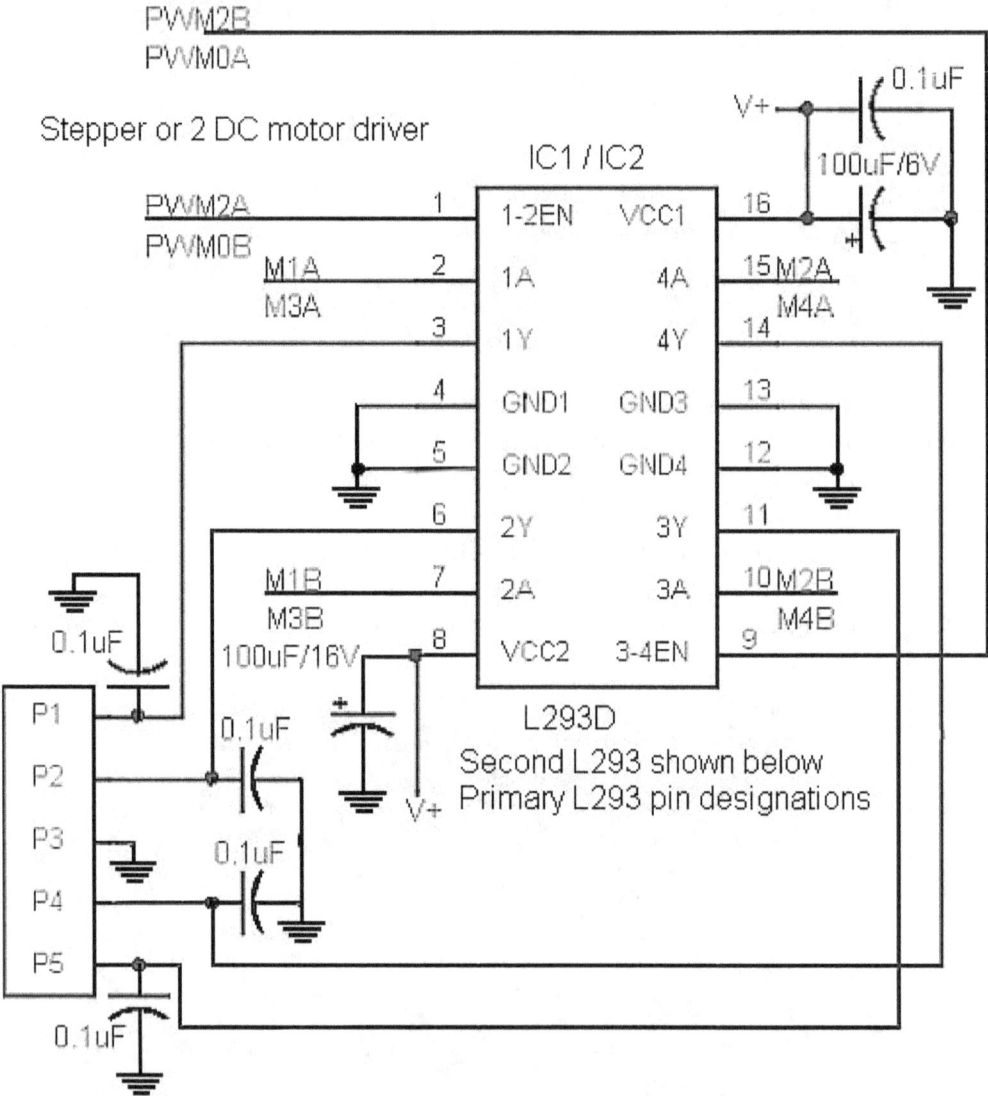

The next schematic diagram shows the connections from the motor control shield going to the Arduino. I have arranged the schematic diagram so that the layout is similar to the layout of the Arduino pins. In this schematic "PWMO" and "PWM2" are PWM (Pulse Width Modulation) pins for the L293's. The pins that are designated PWM1A and PWMB are used for the two optional servo connections.

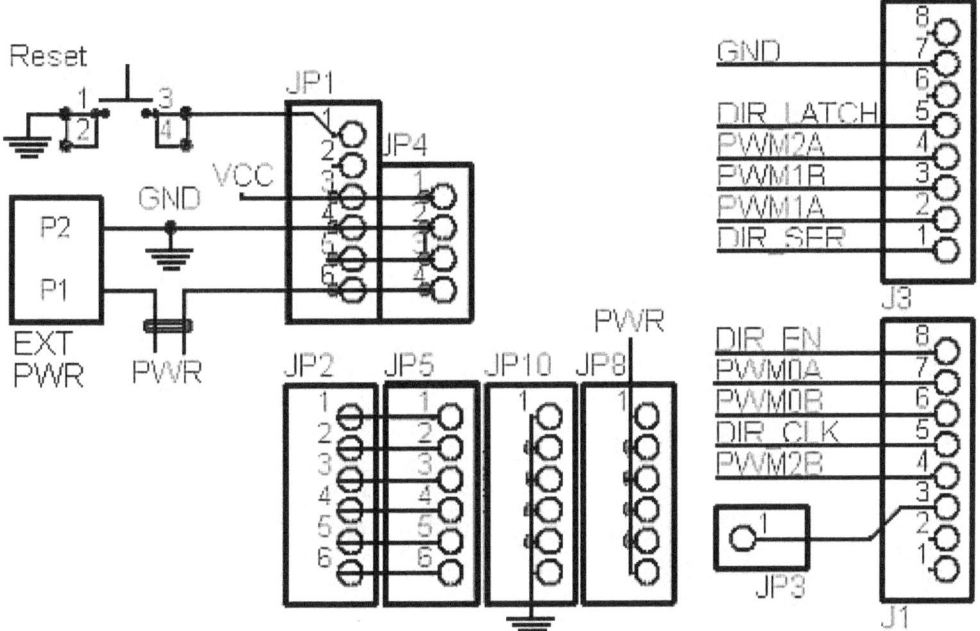

Now for the final additional schematic, we have the 74HC595 eight bit shift register. Rather than use five of the Arduino pins for each L293 IC, the designers of this shield used a shift register to control the motors. Not only did they use a shift register, but the motors that are being controlled do not match up with the pin numbers of the shift register. The additional shift register pins are "SER" as in Serial Data, "CLK" as in Serial Clock, "LATCH" as in Latch the data, and "EN" as in Enable the motors.

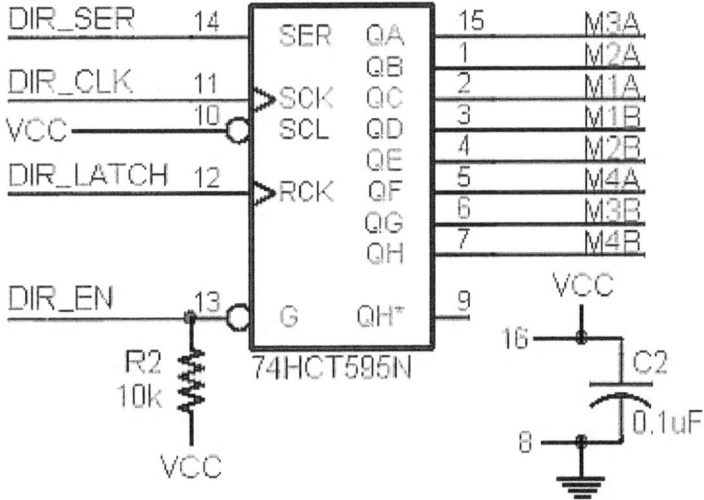

As you might guess this shift register arrangement makes using this motor controller a little bit tricky. Most people will just use the "AdaFruit" driver to make this shield easier to work with. These drivers are found online in the zip file that is called "adafruit-Adafruit-Motor-Shield-library-4bd21ca.zip". It will need to be unzipped into the "Arduino Libraries" folder. I had to rename the unzipped directory to make it compatible with the Arduino software as can be seen in the picture below. If this is not done you will get an error message or the driver will not work.

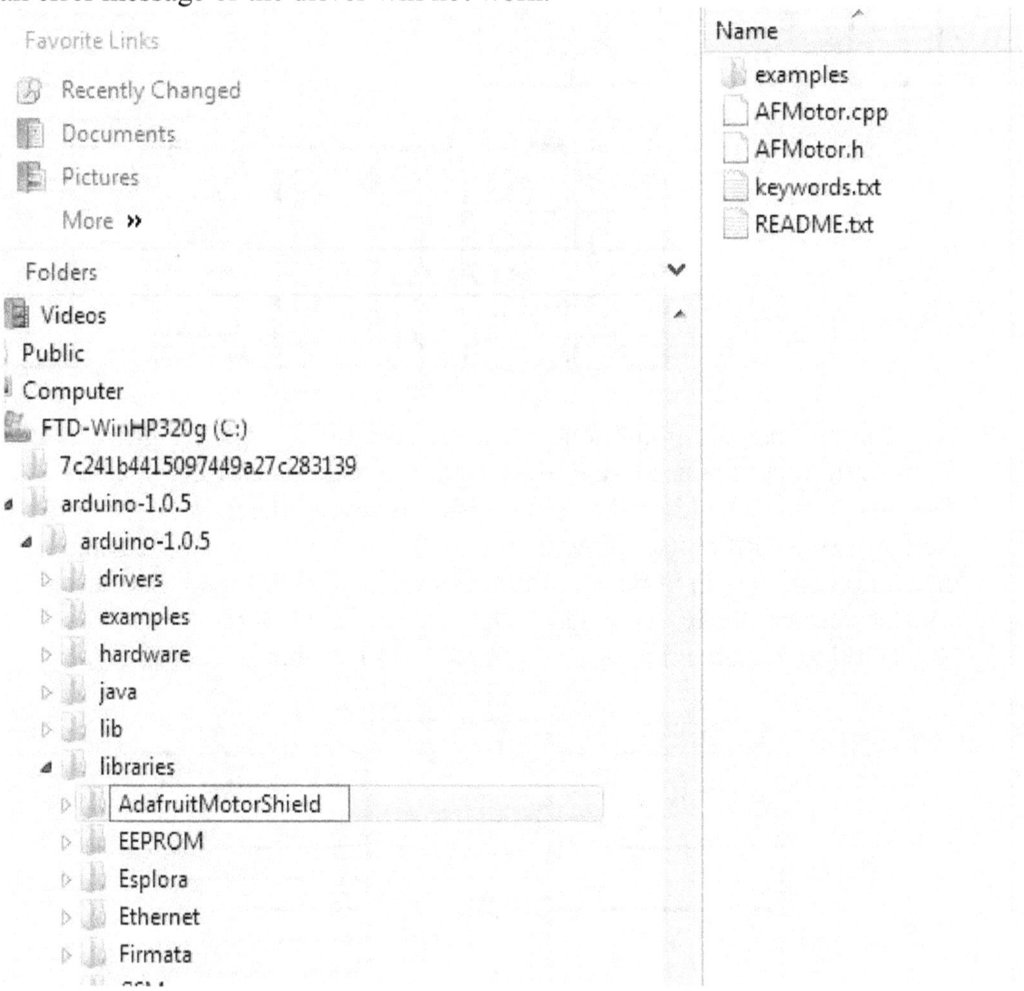

To create a sketch or program to control a motor with this driver software you have to do several steps correctly. To use the Adafruit motor driver in a program you first need to add the following line near the start of the sketch:
#include <AFMotor.h>

Then you need to define the motor that you will be controlling with the next line of code. You will have to do this once for each motor that you will use in your sketch. Each time you define a motor you must have a different name for that motor.
AF_DCMotor motorname(portnumber, frequency)

Here are the parameters:
Motorname - what you will be calling this motor
Portnumber - selects which motor channel (1-4) this motor is connected to
Frequency - selects the PWM frequency.

PWM Frequencies that are available are:
MOTOR12_64KHZ
MOTOR12_8KHZ
MOTOR12_2KHZ
MOTOR12_1KHZ

Here is an example of the code to define a motor:
AF_DCMotor motorleft(1, MOTOR12_64KHZ);
// define a motor called motorleft on channel 1 with 64KHz PWM
From this point on this motor will be defined as "motorleft".

The next command sets the speed of the motor.
setSpeed(speed)
Here are the parameters:
Speed - Valid values for "speed" are between 0 and 255 with 0 being off and 255 is full speed

Here is an example of how to set the motor speed:
motorleft.setSpeed(200); // Set motor left to 4/5 of full speed
DC Motor performance is usually not very linear, the actual RPM will usually not be exactly proportional to the selected speed.

The next command turns the motor on, off, forward and reverse.
run(cmd)
Here are its parameters:
cmd - the desired on/off value for the motor
The values for cmd are:
FORWARD - run forward (actual direction of rotation depends on wiring)
BACKWARD - run backwards (the opposite direction from FORWARD)
RELEASE – Turn off the motor. The motor shield does have any motor stopping circuit, so the motor may spin down slowly.

Next we need to remove the old Robo-raptor controller board and then add the Arduino with the motor controller instead. Here is a diagram showing how I connected the four motors to the controller. I used some 24 gauge telephone wires about 1 inch long to plug into the existing motor connectors on one end and then the other end goes under the screw terminals.

The robo-raptor's motors made a whining noise but they did not move until I added a nine volt power source to the Arduino via its external power connector. This then made more power available to the motor shield. You can use an AC adapter or you can use six "AA" cells to power your raptor.

You might want to eventually connect the LED's in the raptors nose to power via some 1K resistors so they light up as well. There are six analog inputs that are available. They can also be used as outputs to animate other features of the raptor. Perhaps you could even figure out how to make it roar better?

The first program is a sketch that demonstrates the Robo-raptor under Arduino control. When its tail is pinched, it will wiggle, raise and lower its head and then walk forward a few steps, back a few steps and then roar. Walking is just a matter of moving the legs forward one at a time. To detect the pinching of the tail, connect the two yellow wires from P8 (The red and green go to motor 1) to A0 and ground. Also connect a 1K resistor from A0 to 5V.

For making a sound the speaker is connected via the yellow and brown wires of P2 to A6 and ground through a 220uF 16 volt capacitor and a jumper wire.

```
// Robo-Raptor Demo
// demonstrates several actions the Robo raptor can take.
// Written December 2013 by Bob Davis

  // Sequence of Motor control commands:
  // wait for tail switch to get started
  // wiggle tail back and forth
  // rotate neck up and down
  // raise and lower head times
  // take a few steps walking forward
  // take a few steps walking backwards
  // make a roaring sound

#include <AFMotor.h>
#include <Servo.h>
// create servo object to control a servo
Servo myservo;
// mororr moves the right leg
// create motor #2, 64KHz pwm
AF_DCMotor motorr(2, MOTOR12_64KHZ);
// motorl moves the left leg
// create motor #4, 64KHz pwm
AF_DCMotor motorl(4, MOTOR12_64KHZ);
// motorw wiggles the tail, head moves too
// create motor #1, 64KHz pwm
AF_DCMotor motorw(1, MOTOR12_64KHZ);
// motorh moves the head up and down
// create motor #3, 64KHz pwm
AF_DCMotor motorh(3, MOTOR12_64KHZ);
char INBYTE;
// Set A6 as an output pin for speaker
int SpkrPin = 19;
// set A5 as neck motor output
int motorn = 18;
int roar;

void setup() {
  motorr.setSpeed(255);    // set the speed to 255/255
  motorl.setSpeed(255);    // set the speed to 255/255
```

```cpp
  // motor w must have reduced power.
  motorw.setSpeed(255);    // set the speed to 155/255
  motorh.setSpeed(255);    // set the speed to 255/255
  pinMode(SpkrPin, OUTPUT);
  pinMode(motorn, OUTPUT);
  // The servo is on pin 9
  myservo.attach(9);
}

void loop() {
  // Wait for tail switch to start demo
  while (analogRead(A0) != '0'){}

  // wiggle tail
  motorw.run(BACKWARD);   // wiggle right
  delay(200); // pause
  motorw.run(RELEASE);    // stopped
  delay(300);
  motorw.run(FORWARD);    // wiggle left
  delay(200); // pause
  motorw.run(RELEASE);    // stopped
  delay(300);
  motorw.run(BACKWARD);   // wiggle right
  delay(200); // pause
  motorw.run(RELEASE);    // stopped
  delay(300);
  motorw.run(FORWARD);    // wiggle left
  delay(200); // pause
  motorw.run(RELEASE);    // stopped
  delay(300);
  motorw.run(BACKWARD);   // wiggle right
  delay(200); // pause
  motorw.run(RELEASE);    // stopped
  delay(300);
  motorw.run(FORWARD);    // wiggle left
  delay(200); // pause
  motorw.run(RELEASE);    // stopped
  delay(300);

  // wiggle neck motor
  digitalWrite(motorn, HIGH);
  delay(1800);
```

```
    digitalWrite(motorn, LOW);
    delay(300);

    // raise and lower head
    motorh.run(FORWARD);    // raise head
    delay(300);  // pause
    motorh.run(BACKWARD);   // lower head
    delay(300);  // pause
    motorh.run(RELEASE);    // stopped
    delay(700);
    motorh.run(FORWARD);    // raise head
    delay(300);  // pause
    motorh.run(BACKWARD);   // lower head
    delay(300);  // pause
    motorh.run(RELEASE);    // stopped
    delay(700);

  // walk straight forward
  // motorw.run(FORWARD);   // wiggle right
  // delay(300);  // pause
    motorr.run(FORWARD);    // right foot forward
    delay(100);  // pause
    motorw.run(BACKWARD);   // wiggle left
    delay(100);  // pause
    motorr.run(BACKWARD);   // right foot backward
    delay(100);  // pause
    motorr.run(RELEASE);    // stop right foot
  // delay(300);  // pause
    motorl.run(FORWARD);    // left foot forward
    delay(100);  // pause
    motorw.run(FORWARD);    // wiggle right
    delay(100);  // pause
    motorl.run(BACKWARD);   // left foot backward
    delay(100);  // pause
    motorl.run(RELEASE);    // stop left foot
  // delay(300);  // pause
    motorr.run(FORWARD);    // right foot forward
    delay(100);  // pause
    motorw.run(BACKWARD);   // wiggle left
    delay(100);  // pause
    motorr.run(BACKWARD);   // right foot backward
    delay(100);  // pause
```

```
    motorr.run(RELEASE);    // stop right foot
//  delay(300); // pause
    motorl.run(FORWARD);    // left foot forward
    delay(100); // pause
    motorw.run(FORWARD);    // wiggle right
    delay(100); // pause
    motorl.run(BACKWARD);   // left foot backward
    delay(100); // pause
    motorl.run(RELEASE);    // stop left foot
//  delay(300); // pause
    motorw.run(RELEASE);    // stopped
    delay(300);

    // walk straight backward
    motorw.run(BACKWARD);   // wiggle left
    delay(300); // pause
    motorr.run(FORWARD);    // right foot forward
    delay(100); // pause
    motorw.run(FORWARD);    // wiggle right
    dclay(100); // pause
    motorr.run(BACKWARD);   // right foot backward
    delay(100); // pause
    motorr.run(RELEASE);    // stop right foot
//  delay(300); // pause
    motorl.run(FORWARD);    // left foot forward
    delay(100); // pause
    motorw.run(BACKWARD);   // wiggle left
    delay(100); // pause
    motorl.run(BACKWARD);   // left foot backward
    delay(100); // pause
    motorl.run(RELEASE);    // stop left foot
//  delay(300); // pause
    motorr.run(FORWARD);    // right foot forward
    delay(100); // pause
    motorw.run(FORWARD);    // wiggle right
    delay(100); // pause
    motorr.run(BACKWARD);   // right foot backward
    delay(100); // pause
    motorr.run(RELEASE);    // stop right foot
//  delay(300); // pause
    motorl.run(FORWARD);    // left foot forward
    delay(100); // pause
```

```
  motorw.run(BACKWARD);   // wiggle left
  delay(100);  // pause
  motorl.run(BACKWARD);   // left foot backward
  delay(100);  // pause
  motorl.run(RELEASE);    // stop left foot
// delay(300);  // pause
  motorw.run(RELEASE);    // stopped
  delay(300);

  //open mouth
  myservo.write(0);
  delay(300);

  // roar
  for (roar=200; roar > 0; roar--) {
    digitalWrite(SpkrPin, HIGH);   // sets the speaker on
    delay(random(10));      // waits for a fraction of a second
    digitalWrite(SpkrPin, LOW);    // sets the speaker off
    delay(random(10));      // waits for a fraction of a second
  }
  //close mouth
  myservo.write(90);
  delay(300);
}
```

You can optionally control the raptors neck motor with a TIP120 or TIP121 power transistor. Here is the schematic diagram. The wires to that motor are colored red and purple. It goes in circles so you do not need to reverse it.

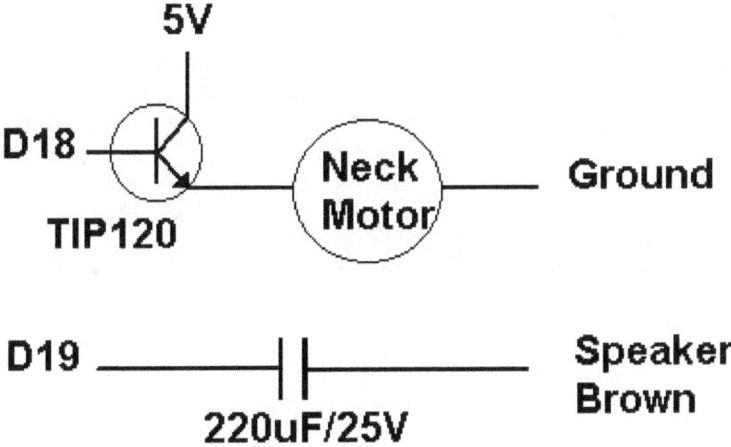

D18 is the same as A5 and D19 is the same ad A6. They are available on the motor control circuit board. I added some breadboard pins so I could easily add circuitry there. There are neck position switches in the raptor to tell you when the neck is up or down. I did not bother with monitoring them.

I also at one point added a servo motor to control the mouth. I had to disconnect the cable to the mouth located on the left side and connect it to the servo. I used a piece of 20 gauge wire to connect it to the servo. The servo is glued in place just in front of the Arduino. The servo is controlled by D9 and should be plugged into "servo 1" on the circuit board. However it only worked properly when it is plugged into "servo 2", so I think there is an error in the circuit board design.

In this picture you can see the servo in the lower left corner, the speaker capacitor at the top center and the power transistor that controls the neck motor is also in the top center of the picture.

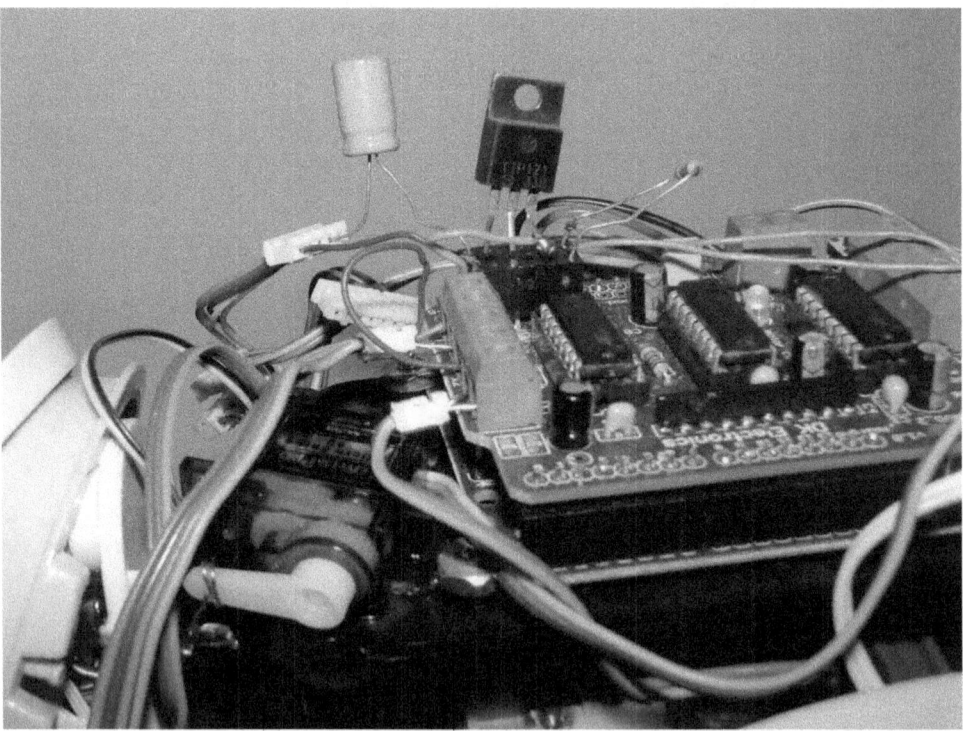

The next program listed here gives you control of the Robo-raptor via serial communications. You can use either the USB cable or the Bluetooth adapter that was explained in an earlier chapter. With Bluetooth control you can even control the Robo-raptor from your cell phone!

```cpp
// Robo-Raptor serial
// responds to serial commands via USB or Bluetooth
// Written December 2013 by Bob Davis
  // Motor control commands
  // b = walk backwards
  // f = walk forward
  // h = raise head
  // l = lower head
  // w = wiggle tail
  // r = roar sound
  // m = open mouth
  // n = wiggle neck
  // D9 is servo 1 labeled as servo 2 on circuit board.

#include <AFMotor.h>
#include <Servo.h>
// create servo object to control a servo
Servo myservo;
// motorr moves right foot
// create motor #2, 64KHz pwm
AF_DCMotor motorr(2, MOTOR12_64KHZ);
// motorl moves left foot
// create motor #4, 64KHz pwm
AF_DCMotor motorl(4, MOTOR34_64KHZ);
// motorw is motor wiggle, head and tail move together
// create motor #1, 64KHz pwm
AF_DCMotor motorw(1, MOTOR12_64KHZ);
// motorh moves the head
// create motor #3, 64KHz pwm
AF_DCMotor motorh(3, MOTOR34_1KHZ);
char INBYTE;
// set A6 as an output pin for speaker
int SpkrPin = 19;
// set A5 as neck motor output
int motorn = 18;
int roar;

void setup() {
  motorr.setSpeed(255);    // set the speed to 255/255
  motorl.setSpeed(255);    // set the speed to 255/255
  motorw.setSpeed(255);    // set the speed to 150/255
```

```
  motorh.setSpeed(255);    // set the speed to 255/255
  Serial.begin(9600);
  pinMode(SpkrPin, OUTPUT);
  pinMode(motorn, OUTPUT);
  // The servo is on pin 9
  myservo.attach(9);
}

void loop() {
  // read next available byte
  INBYTE = Serial.read();
  if (INBYTE == 'n'){
  // wiggle neck motor
  digitalWrite(motorn, HIGH);
  delay(300);
  digitalWrite(motorn, LOW);
  delay(300);
  }
  if (INBYTE == 'm'){
  // open mouth
  myservo.write(0);
  delay(300);
  myservo.write(90);
  delay(300);
  }
  if (INBYTE == 'b'){
  // walk straight backward
  motorw.run(BACKWARD);    // wiggle left
  delay(300);  // pause
  motorr.run(FORWARD);     // right foot forward
  delay(100);  // pause
  motorw.run(FORWARD);     // wiggle right
  delay(100);   // pause
  motorr.run(BACKWARD);    // right foot backward
  delay(100);  // pause
  motorr.run(RELEASE);    // stop right foot
  motorl.run(FORWARD);     // left foot forward
  delay(100);  // pause
  motorw.run(BACKWARD);    // wiggle left
  delay(100);   // pause
  motorl.run(BACKWARD);    // left foot backward
  delay(100);  // pause
```

```
motorl.run(RELEASE);      // stop left foot
motorr.run(FORWARD);      // right foot forward
delay(100);  // pause
motorw.run(FORWARD);      // wiggle right
delay(100);  // pause
motorr.run(BACKWARD);     // right foot backward
delay(100);  // pause
motorr.run(RELEASE);      // stop right foot
motorl.run(FORWARD);      // left foot forward
delay(100);  // pause
motorw.run(BACKWARD);     // wiggle left
delay(100);  // pause
motorl.run(BACKWARD);     // left foot backward
delay(100);  // pause
motorl.run(RELEASE);      // stop left foot
motorw.run(RELEASE);      // stopped
delay(300);
}
if (INBYTE == 'f'){
// walk straight forward
motorr.run(FORWARD);      // right foot forward
delay(100);  // pause
motorw.run(BACKWARD);     // wiggle left
delay(100);  // pause
motorr.run(BACKWARD);     // right foot backward
delay(100);  // pause
motorr.run(RELEASE);      // stop right foot
motorl.run(FORWARD);      // left foot forward
delay(100);  // pause
motorw.run(FORWARD);      // wiggle right
delay(100);  // pause
motorl.run(BACKWARD);     // left foot backward
delay(100);  // pause
motorl.run(RELEASE);      // stop left foot
motorr.run(FORWARD);      // right foot forward
delay(100);  // pause
motorw.run(BACKWARD);     // wiggle left
delay(100);  // pause
motorr.run(BACKWARD);     // right foot backward
delay(100);  // pause
motorr.run(RELEASE);      // stop right foot
motorl.run(FORWARD);      // left foot forward
```

```
  delay(100);  // pause
  motorw.run(FORWARD);    // wiggle right
  delay(100);  // pause
  motorl.run(BACKWARD);   // left foot backward
  delay(100);  // pause
  motorl.run(RELEASE);    // stop left foot
  motorw.run(RELEASE);    // stopped
  delay(300);
}
if (INBYTE == 'h'){
// raise head
  motorh.run(FORWARD);    // raise head
  delay(800);  // pause
  motorh.run(RELEASE);    // stopped
  delay(300);
}
if (INBYTE == 'l'){
// lower head
  motorh.run(BACKWARD);   // lower head
  delay(300);  // pause
  motorh.run(RELEASE);    // stopped
  delay(300);
}
if (INBYTE == 'w'){
// wiggle tail
  motorw.run(FORWARD);    // wiggle right
  delay(300);  // pause
  motorw.run(RELEASE);    // stopped
  delay(300);
  motorw.run(BACKWARD);   // wiggle left
  delay(300);  // pause
  motorw.run(RELEASE);    // stopped
  delay(300);
}
if (INBYTE == 'r'){   // roar
for (roar=200; roar > 0; roar--) {
  digitalWrite(SpkrPin, HIGH);  // sets the speaker on
  delay(random(10));       // waits for a fraction of a second
  digitalWrite(SpkrPin, LOW);   // sets the speaker off
  delay(random(10));       // waits for a fraction of a second
 }
}
```

}

Here is how to run the Robo-raptor with an Infrared Remote control.

You can possibly control the Robo-raptor with infrared, but there are several obstacles you have to overcome first. The number one problem is that the IR remote library is not compatible with the Adafruit motor controller driver. If you load the IR driver first when you load the motor driver the IR starts returning garbage. If you load the IR driver second the motors stop working. There are three possible solutions. Make your own motor controller, write your own motor controller software of fix the Adafruit driver problem. The last solution is the easiest one to do.

Tests with a meter determined that the problem was that the motor enables were turned off. However we know what the pins are that those enables are connected to. All we need to do is to force those pins high and hence force the motors to be on.

The next problem is to connect the IR receiver to the Arduino. D2 is not used but it does not come out to a pin to make it easy to attach the IR receiver to. I have added a six pin header to pins D0 to D5 of the motor controller so we can access D2 as well as D0 and D1 for the serial/Bluetooth control. By adding that, D2 can now be easily connected to the IR receiver.

The next step is to define what the buttons will do on the remote control. The top buttons on the remote do 'composite' operations that require multiple things to happen together. The bottom buttons on the remote, numbers one through nine, are set up so that each button controls an individual motor one button for forward and one button for reverse. Here is a chart showing the new remote control definitions.

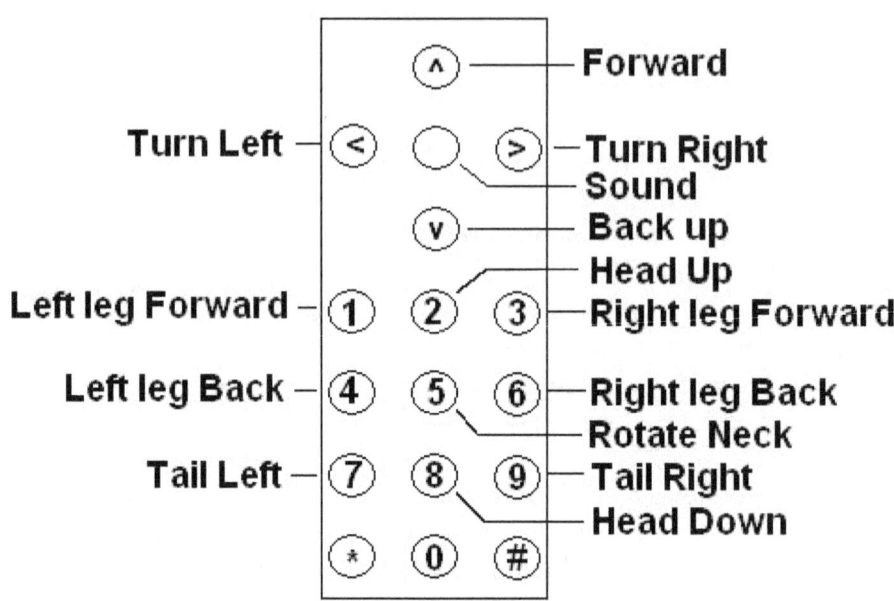

Here is the RoboRaptor sketch for IR control of the robot.

```
/******************************************
Robotics-RoboRaptor IR Remote control
Demonstrates the RoboRaptor under IR control
Written by Bob Davis February 2014
  // Motor control commands
  // ^ = walk forward
  // v = walk backwards
  // > = walk to the right
  // < = walk to the left
  // 1, 4 left leg front, back
  // 3, 6 right leg front, back
  // 2 = raise head
  // 5 = wiggle neck
  // 7 = wiggle tail left
  // 8 = lower head
  // 9 = wiggle tail right
  // OK = roaring sound
*****************************************/

#include <AFMotor.h>
// motorr #2 moves right foot
AF_DCMotor motorr(2, MOTOR12_64KHZ);
```

```
// motorl #4 moves left foot
AF_DCMotor motorl(4, MOTOR34_64KHZ);
// motorw #1 wiggle, head and tail move together
AF_DCMotor motorw(1, MOTOR12_64KHZ);
// motorh #3 moves the head up and down
AF_DCMotor motorh(3, MOTOR34_64KHZ);
// set A6 as an output pin for speaker
int SpkrPin = 19;
// set A5 as neck motor output
int motorn = 18;
int roar;
// set up to override the motor enables
int D3 = 3;
int D5 = 5;
int D6 = 6;
int D11 = 11;
// IR setup must load this AFTER AFMotor!
#include <IRremote.h>
int RECV_PIN = 2;
IRrecv irrecv(RECV_PIN);
decode_results results;

void setup() {
  Serial.begin(9600);
  motorr.setSpeed(255);  // set the speed to 255/255
  motorl.setSpeed(255);  // set the speed to 255/255
  motorw.setSpeed(255);  // set the speed to 150/255
  motorh.setSpeed(255);  // set the speed to 255/255
  // IRremote causes motors to turn off so I will force them on
  pinMode(D3, OUTPUT);
  pinMode(D5, OUTPUT);
  pinMode(D6, OUTPUT);
  pinMode(D11, OUTPUT);

  pinMode(SpkrPin, OUTPUT);
  pinMode(motorn, OUTPUT);
  irrecv.enableIRIn();  // Start the receiver
}

void loop() {
  digitalWrite(D3, HIGH);
  digitalWrite(D5, HIGH);
```

```
digitalWrite(D6, HIGH);
digitalWrite(D11, HIGH);
if (irrecv.decode(&results)) {

  if (results.value==16738455) {
    Serial.println("1 key");
// left foot forward
motorl.run(FORWARD);
delay(100);  // pause
motorl.run(RELEASE);    // stopped
delay(100);
  }

  if (results.value==16750695) {
    Serial.println("2 key");
// raise head
motorh.run(FORWARD);    // raise head
delay(300);  // pause
motorh.run(RELEASE);    // stopped
delay(100);
  }

  if (results.value==16756815) {
    Serial.println("3 key");
// right foot forward
motorr.run(FORWARD);
delay(100);  // pause
motorr.run(RELEASE);    // stopped
delay(100);
  }

  if (results.value==16724175) {
    Serial.println("4 key");
// left foot back
motorl.run(BACKWARD);
delay(100);  // pause
motorl.run(RELEASE);    // stopped
delay(100);
  }

  if (results.value==16718055) {
    Serial.println("5 key");
```

```
// wiggle neck motor
digitalWrite(motorn, HIGH);
delay(300);
digitalWrite(motorn, LOW);
delay(100);
  }

  if (results.value==16743045) {
    Serial.println("6 key");
// right leg back
motorr.run(BACKWARD);
delay(100);  // pause
motorr.run(RELEASE);     // stopped
delay(100);
  }
  if (results.value==16716015) {
    Serial.println("7 key");
// wiggle tail left
motorw.run(BACKWARD);    // wiggle left
delay(300);  // pause
motorw.run(RELEASE);     // stopped
delay(100);
  }

  if (results.value==16726215) {
    Serial.println("8 key");
// lower head
motorh.run(BACKWARD);    // lower head
delay(300);  // pause
motorh.run(RELEASE);     // stopped
delay(100);
  }

  if (results.value==16734885) {
    Serial.println("9 key");
// wiggle tail right
motorw.run(FORWARD);     // wiggle right
delay(300);  // pause
motorw.run(RELEASE);     // stopped
delay(100);
  }
```

```
if (results.value==16730805) {
  Serial.println("0 key");
}

if (results.value==16728765) {
  Serial.println("star key");
}

if (results.value==16732845) {
  Serial.println("pound key");
}
if (results.value==16712445) {
  Serial.println("OK key");
// roaring sound
for (roar=200; roar > 0; roar--) {
  digitalWrite(SpkrPin, HIGH);   // sets the speaker on
  delay(random(10));       // waits for a fraction of a second
  digitalWrite(SpkrPin, LOW);   // sets the speaker off
  delay(random(10));       // waits for a fraction of a second
  }
}

if (results.value==16736925) {
  Serial.println("forward key");
  // Go forward
motorr.run(FORWARD);     // right foot forward
delay(100); // pause
motorw.run(BACKWARD);     // wiggle left
delay(100); // pause
motorr.run(BACKWARD);     // right foot backward
delay(100); // pause
motorr.run(RELEASE);    // stop right foot
motorl.run(FORWARD);     // left foot forward
delay(100); // pause
motorw.run(FORWARD);     // wiggle right
delay(100); // pause
motorl.run(BACKWARD);     // left foot backward
delay(100); // pause
motorl.run(RELEASE);    // stop left foot
motorr.run(FORWARD);     // right foot forward
delay(100); // pause
motorw.run(BACKWARD);     // wiggle left
```

```
delay(100); // pause
motorr.run(BACKWARD);   // right foot backward
delay(100);  // pause
motorr.run(RELEASE);    // stop right foot
motorl.run(FORWARD);    // left foot forward
delay(100);  // pause
motorw.run(FORWARD);    // wiggle right
delay(100);  // pause
motorl.run(BACKWARD);   // left foot backward
delay(100);  // pause
motorl.run(RELEASE);    // stop left foot
motorw.run(RELEASE);    // stopped
delay(300);
  }

  if (results.value==16761405) {
    Serial.println("right key");
    // Turn right
motorr.run(FORWARD);    // right foot forward
delay(100); // pause
motorw.run(BACKWARD);   // wiggle left
delay(100);  // pause
motorr.run(BACKWARD);   // right foot backward
delay(100);  // pause
motorr.run(RELEASE);    // stop right foot
motorl.run(FORWARD);    // left foot forward
delay(100);  // pause
motorw.run(FORWARD);    // wiggle right
delay(400);  // pause
motorl.run(BACKWARD);   // left foot backward
delay(100);  // pause
motorl.run(RELEASE);    // stop left foot
motorw.run(RELEASE);    // stop wiggle
  }

  if (results.value==16754775) {
    Serial.println("back key");
    // Go Backwards
motorw.run(BACKWARD);   // wiggle left
delay(300);  // pause
motorr.run(FORWARD);    // right foot forward
delay(100);  // pause
```

```
motorw.run(FORWARD);     // wiggle right
delay(100);  // pause
motorr.run(BACKWARD);    // right foot backward
delay(100);  // pause
motorr.run(RELEASE);     // stop right foot
motorl.run(FORWARD);     // left foot forward
delay(100);  // pause
motorw.run(BACKWARD);    // wiggle left
delay(100);  // pause
motorl.run(BACKWARD);    // left foot backward
delay(100);  // pause
motorl.run(RELEASE);     // stop left foot
motorr.run(FORWARD);     // right foot forward
delay(100);  // pause
motorw.run(FORWARD);     // wiggle right
delay(100);  // pause
motorr.run(BACKWARD);    // right foot backward
delay(100);  // pause
motorr.run(RELEASE);     // stop right foot
motorl.run(FORWARD);     // left foot forward
delay(100);  // pause
motorw.run(BACKWARD);    // wiggle left
delay(100);  // pause
motorl.run(BACKWARD);    // left foot backward
delay(100);  // pause
motorl.run(RELEASE);     // stop left foot
motorw.run(RELEASE);     // stopped
delay(300);
  }

  if (results.value==16720605) {
    Serial.println("left key");
    // Turn left
motorr.run(FORWARD);     // right foot forward
delay(100);  // pause
motorw.run(BACKWARD);    // wiggle left
delay(400);  // pause
motorr.run(BACKWARD);    // right foot backward
delay(100);  // pause
motorr.run(RELEASE);     // stop right foot
motorl.run(FORWARD);     // left foot forward
delay(100);  // pause
```

```
    motorw.run(FORWARD);      // wiggle right
    delay(100);   // pause
    motorl.run(BACKWARD);     // left foot backward
    delay(100);   // pause
    motorl.run(RELEASE);      // stop left foot
    motorw.run(RELEASE);      // stop wiggle
      }

// Next two lines are for troubleshooting so you can see the codes
    Serial.println(results.value);
//    dump(&results);
   irrecv.resume(); // Receive the next value
   }
 }
```

Well hopefully you have had a lot of fun learning about some of the building blocks of robotics. This book is only meant to start you on your way to a robotics adventure. There are many more subjects that are not covered in this book, such as building competitive robots, life sized robots, and many others.

# Bibliography

**Programming Arduino**
Getting Started With Sketches
By Simon Mark
Copyright 2012 by the McGraw-Hill Companies

This book gives a thorough explanation of the programming code for the Arduino. However the projects in the book are very basic. It does cover LCD's and Ethernet adapters.

**Getting Started with Arduino**
By Massimo Banzi
Copyright 2011 Massimo Banzi

This author is a co-founder of the Arduino. This book has a quick reference to the programming code and some simple projects.

**Arduino Cookbook**
by Michael Margolis
Copyright © 2011 Michael Margolis and Nicholas Weldin. All rights reserved.
Published by O'Reilly Media, Inc., 1005 Gravenstein Highway North, Sebastopol, CA.

This book has lots of great projects, with a very good explanation for every project.

**Practical Arduino: Cool Projects for Open Source Hardware**
Copyright © 2009 by Jonathan Oxer and Hugh Blemings

Page 197 of this book has how to supercharge the Analog to Digital converter for faster sampling rates in oscilloscopes and logic analyzers.

www.ingramcontent.com/pod-product-compliance
Lightning Source LLC
Chambersburg PA
CBHW080307180526
45167CB00006B/2707